AF686806

This work: "Assessing Recent Soil Erosion Rates through the Use of Beryllium-7 (Be-7)", was originally published by Springer Nature and is being sold pursuant to a Creative Commons license permitting commercial use. All rights not granted by the work's license are retained by the author or authors.

Cover Image Credit/Copyright Attribution: IIerlok_Xolms/Shutterstock

Lionel Mabit · William Blake
Editors

Assessing Recent Soil Erosion Rates through the Use of Beryllium-7 (Be-7)

Editors
Lionel Mabit
Soil and Water Management
and Crop Nutrition Subprogramme
Joint FAO/IAEA Division of Nuclear
Techniques in Food and Agriculture
International Atomic Energy Agency
Vienna, Austria

William Blake
School of Geography, Earth
and Environmental Sciences
University of Plymouth
Plymouth, UK

https://doi.org/10.1007/978-3-030-10982-0

© International Atomic Energy Agency (IAEA) 2019. This book is an open access publication.
Open Access provided with a grant from the International Atomic Energy Agency.
The opinions expressed in this publication are those of the authors/editors and do not necessarily reflect the views of the International Atomic Energy Agency, its Board of Directors, or the countries they represent.
Open Access This book is licensed under the terms of the Creative Commons Attribution 3.0 IGO license (https://creativecommons.org/licenses/by/3.0/igo/), which permits use, sharing, adaptation, distribution and reproduction in any medium or format, as long as you give appropriate credit to the International Atomic Energy Agency, provide a link to the Creative Commons license and indicate if changes were made.
The use of the International Atomic Energy Agency's name for any purpose other than for attribution, and the use of the International Atomic Energy Agency's logo, shall be subject to a separate written license agreement between the International Atomic Energy Agency and the user and is not authorized as part of this CC-IGO license. Note that the link provided above includes additional terms and conditions of the license.
The images or other third party material in this book are included in the book's Creative Commons license, unless indicated otherwise in a credit line to the material. If material is not included in the book's Creative Commons license and your intended use is not permitted by statutory regulation or exceeds the permitted use, you will need to obtain permission directly from the copyright holder.
The use of general descriptive names, registered names, trademarks, service marks, etc. in this publication does not imply, even in the absence of a specific statement, that such names are exempt from the relevant protective laws and regulations and therefore free for general use.
The publisher, the authors, and the editors are safe to assume that the advice and information in this book are believed to be true and accurate at the date of publication. Neither the publisher nor the authors or the editors give a warranty, express or implied, with respect to the material contained herein or for any errors or omissions that may have been made. The publisher remains neutral with regard to jurisdictional claims in published maps and institutional affiliations.

Preface

The Soil and Water Management & Crop Nutrition (SWMCN) Subprogramme of the Joint FAO/IAEA Division of Nuclear Techniques in Food and Agriculture supports IAEA Member States and FAO Member Countries in the use of fallout radionuclides (FRNs) techniques for assessing soil redistribution patterns and magnitudes to reduce the impact of soil erosion on agro-ecosystems.

From all available FRNs (e.g. ^{137}Cs, $^{210}Pb_{ex}$, $^{239+240}Pu$), Beryllium-7 (^{7}Be) is the only soil tracer that permits the gathering of short-term soil redistribution information. This natural cosmogenic isotope was used in this context for the first time at the end of the 1990s. Developments over the past decades now permit investigation of erosion processes not only during short extreme weather events but also over extended periods of up to several months.

The SWMCN Subprogramme has actively sought to refine the use of ^{7}Be in close collaboration with the School of Geography, Earth and Environmental Sciences at the University of Plymouth, UK, and with the help of scientists from developed and developing countries. This book reflects the latest developments obtained and the state of the art related to the use of this natural tracer of soil and sediment.

This handbook presents the foundation of the ^{7}Be method and provides guidelines for the practical applications of basic and advanced approaches of this technique. It assists scientists, technicians and students to assess the impact of soil erosion events and further support the development of land management and climate-smart agriculture policies for future sustainable food security.

Vienna, Austria — Lionel Mabit
Plymouth, UK — William Blake

Contents

Chapter 1
The Use of Be-7 as a Soil and Sediment Tracer

A. Taylor, W. H. Blake, A. R. Iurian, G. E. Millward and L. Mabit

1.1 Origin and Agro-Environmental Behaviour of ^{7}Be

1.1.1 Atmospheric Production of ^{7}Be

Beryllium-7 (^{7}Be) ($T_{1/2}$ = 53.3 days) is a cosmogenic fallout radionuclide (FRN) produced in the upper atmosphere by cosmic ray spallation of nitrogen and oxygen. As reported by Kaste et al. (2002), rates of production are dependent upon solar activity with greater production in the stratosphere than the troposphere and generally increased production occurring at higher latitudes owing to cosmic ray deflection towards the poles.

Following its formation, ^{7}Be becomes associated with aerosols and its flux is then controlled by complex atmospheric transport processes, which display seasonal variations largely linked to atmospheric mixing and precipitation patterns. During

A. Taylor (✉) · G. E. Millward
Consolidated Radioisotope Facility (CoRiF), University of Plymouth, Plymouth, UK
e-mail: alex.taylor@plymouth.ac.uk

G. E. Millward
e-mail: G.Millward@plymouth.ac.uk

W. H. Blake
School of Geography, Earth and Environmental Sciences, University of Plymouth, Plymouth, UK
e-mail: william.blake@plymouth.ac.uk

A. R. Iurian
Terrestrial Environment Laboratory, IAEA Laboratories Seibersdorf, Seibersdorf, Austria
e-mail: A.Iurian@iaea.org

L. Mabit
Soil and Water Management and Crop Nutrition Subprogramme, Joint FAO/IAEA Division of Nuclear Techniques in Food and Agriculture, International Atomic Energy Agency, Vienna, Austria
e-mail: L.Mabit@iaea.org

© International Atomic Energy Agency (IAEA) 2019
L. Mabit and W. Blake (eds.), *Assessing Recent Soil Erosion Rates through the Use of Beryllium-7 (Be-7)*, https://doi.org/10.1007/978-3-030-10982-0_1

the spring months at mid latitudes, increased stratosphere-troposphere exchange can occur during folding of the tropopause leading to higher concentrations of ^{7}Be in the troposphere (Kaste et al. 2002). Once in the troposphere, ^{7}Be is subjected to vertical mixing through convective circulation, which is likely to increase during warmer months and serves to transport ^{7}Be enriched air to the lower troposphere, increasing its availability for precipitation scavenging. Given that precipitation scavenging is the main mechanism for the removal of ^{7}Be from the atmosphere, seasonal and latitudinal climatic conditions exert a strong influence upon atmospheric concentrations (Kusmierczyk-Michulec et al. 2015).

1.1.2 ^{7}Be Fallout

Wet deposition governs ^{7}Be delivery to the Earth's surface while dry deposition only accounts for approximately 10% of the overall fallout (Kaste et al. 2002). Activity concentration of ^{7}Be in rainwater (Bq L^{-1}) is conditional on its availability for scavenging and is influenced by a number of factors including rainfall rates and volumes (Ioannidou and Papastefanou 2006).

The term 'washout' refers to situations when rainfall scavenging exceeds the availability of ^{7}Be in the atmosphere leading to a reduction in ^{7}Be rainwater activity concentration across an event and following high magnitude or prolonged rainfall events. In contrast, greater activity concentrations during low magnitude events can be related to efficient scavenging of ^{7}Be by fine droplets where there is an abundance of ^{7}Be-bearing aerosols (Ioannidou and Papastefanou 2006). The processes of atmospheric circulation and depletion not only affect the activity concentrations of ^{7}Be on a temporal scale, but are also likely to increase its spatial variability (Taylor et al. 2016). ^{7}Be deposition (Bq m^{-2}) is significantly correlated to the amount of rainfall received on an event basis and, therefore, seasonal depositional fluxes reflect monthly rainfall volumes (Doering and Akber 2008).

1.1.3 ^{7}Be Sorption Behaviour in Soils and Sediments

Upon deposition and infiltration into the soil surface, ^{7}Be is known to be rapidly adsorbed to fine soil particles (Fig. 1.1), which is supported by laboratory batch studies, high partition coefficients ($K_d > 10^5$) in aquatic systems and shallow depth distributions displayed in a range of soil types (Taylor et al. 2013). ^{7}Be shows preferential adsorption to fine sediment fractions (Taylor et al. 2014), which are likely to be readily mobilised in hillslope systems. In solutions of pH 5–6 and in the absence of humic acid, ^{7}Be is likely to be present as Be^{2+} and the hydrolysed species, $BeOH^+$; both of these forms are highly reactive, resulting in rapid sorption rates (Kaste et al. 2002). The rapid adsorption together with a short half-life (the latter precludes the effects of diffusion and bioturbation over time) leads to distributions typically extend-

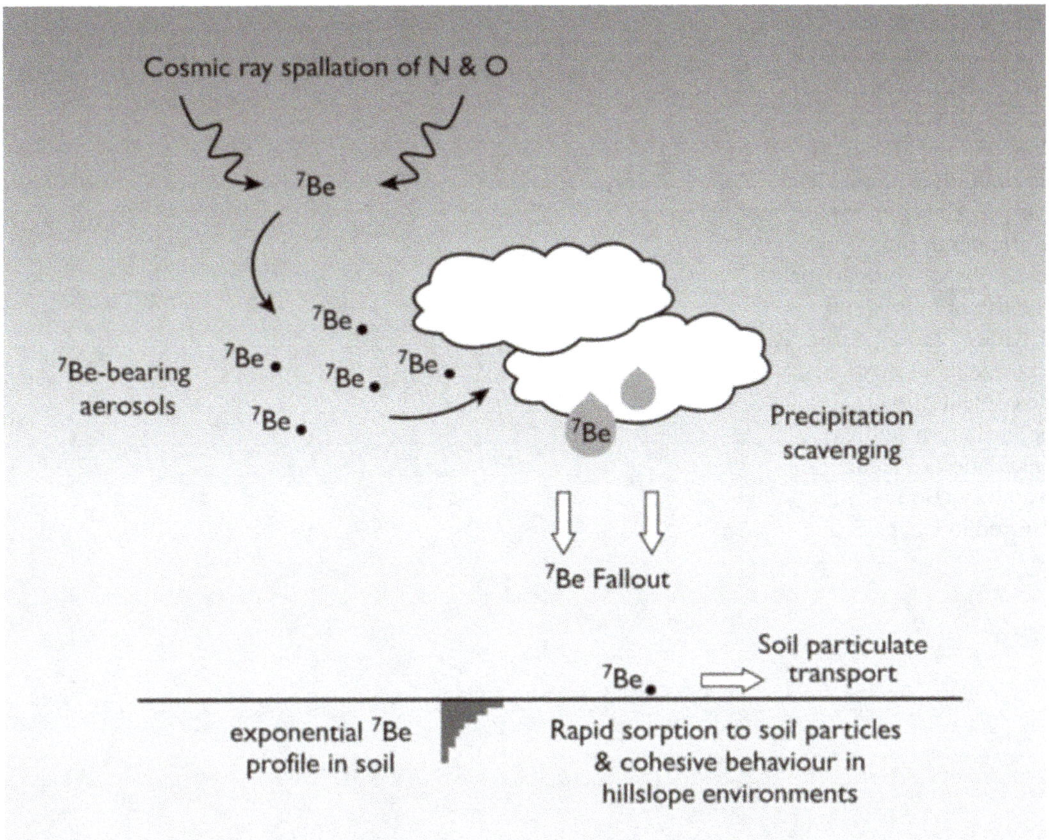

Fig. 1.1 Schematic diagram of ^{7}Be production and fallout. ^{7}Be is produced by cosmic ray spallation in the upper atmosphere and becomes associated with aerosols. ^{7}Be-bearing aerosols are scavenged by precipitation and, consequently, wet deposition is the dominant pathway to the Earth's surface. ^{7}Be is rapidly adsorbed to soil particles and remains cohesive under oxic field conditions

ing to around 20 mm below the soil surface and an exponential decrease with depth (Fig. 1.2). Where soils remain aerated and free from waterlogged conditions, as is the case in most agricultural field sites, ^{7}Be is likely to remain adsorbed to soil particles (Taylor et al. 2012). Both characteristics (i.e. rapid adsorption and cohesive behaviour) are key prerequisites for ^{7}Be application as a soil redistribution tracer.

1.2 The Use of ^{7}Be as a Soil Redistribution Tracer

The application of FRN tracers (e.g. ^{7}Be, ^{137}Cs, ^{210}Pb$_{ex}$, Pu) provides distinct and additional advantages over traditional soil monitoring techniques by enabling retrospective estimates of soil redistribution from relatively few site visits. Through carefully planned sampling programmes, high spatial resolution estimates of soil redistribution can be acquired, which are unlikely to be achieved using conventional methods (Walling et al. 1999). In this regard, the use of nuclear and iso-

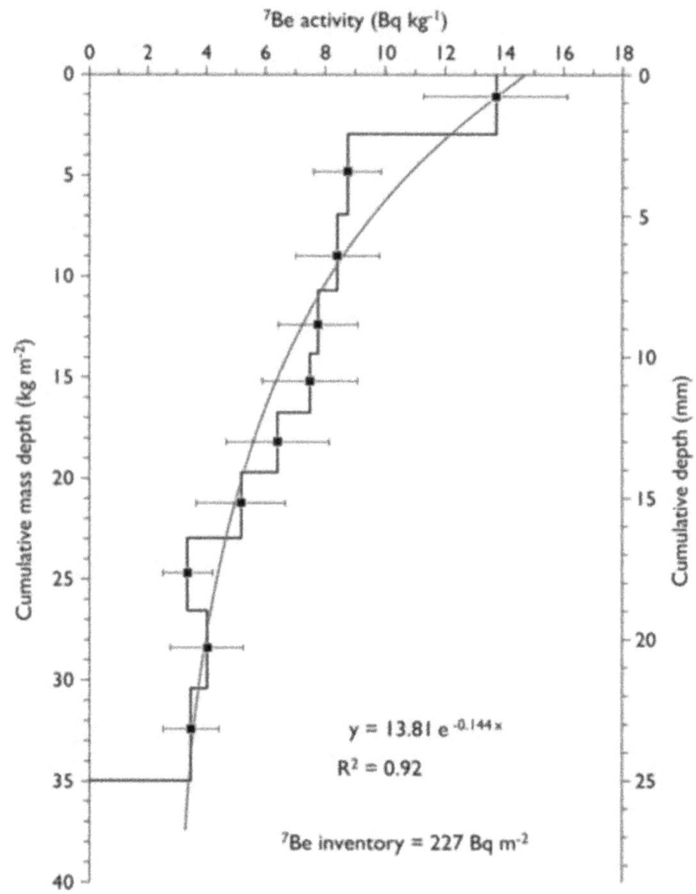

Fig. 1.2 Example of ^{7}Be depth distribution in a clay loam soil in Seibersdorf, Austria (Iurian et al. 2013). Note that the ^{7}Be depth distribution is plotted as a mass depth (kg m^{-2}) and the metric depth (mm) is shown here only for comparison. The use of mass depth provides a more precise and accurate measure of depth in the context of soil redistribution studies. Calculation of areal inventory (Bq m^{-2}) is covered in Chap. 2

topic techniques enables rigorous assessment of soil conservation measures and, if applied over extended time periods, can help to determine the sustainability of agricultural systems.

FRN tracing techniques have been commonly applied to assess soil redistribution at the hillslope scale using ^{137}Cs ($T_{1/2} = 30.2$ years) and ^{210}Pb ($T_{1/2} = 22.3$ years). The half-life of these radionuclides enables redistribution estimates relating to medium-term (i.e. decadal) timescales, providing important information with regard to soil loss or gain in an historical context. However, the application of these decadal-scale radionuclides does not allow for determination of soil redistribution as a result of recent changes in land use or recent rainfall events. This is an important consideration given the need for assessment of soil conservation measures with increasing recognition of soil as a finite global resource. The short-lived ^{7}Be is, therefore, complementary to the other FRN tracers in enabling estimates of soil redistribution across much shorter timescales such as rainfall events or wet seasons.

The premise for using ^{7}Be as a soil redistribution tracer is that the tracer will be rapidly adsorbed to soil particles upon fallout, thus establishing an inventory (^{7}Be activity per unit area) (Bq m^{-2}) at a sampling site. A key assumption of the technique is that ^{7}Be will remain adsorbed to soil particles across the timescale of the study and, therefore, any changes in inventory (i.e. depletion or gain) at a sampling point can only occur as a result of fallout, radioactive decay and/or soil redistribution processes. The ^{7}Be inventory at a site, which experiences soil redistribution, can be compared directly to the ^{7}Be inventory at a stable location, referred to as a reference site, where there is no loss or gain in inventory as a result of soil redistribution. The inventory at a reference site is established only through ^{7}Be atmospheric fallout and, given that this is predominantly delivered by precipitation, reference sites are typically level areas which retain the fallout without loss except for radioactive decay. Alongside the important tracer assumptions of rapid adsorption upon fallout and cohesive behaviour during the period of study, it is also crucial that the reference site receives the same fallout deposition as the eroding area of interest, which requires both sites to be in close proximity. In addition to this, it is also assumed that the fallout received at the study site will be uniform so any differences in inventory across the study area can be attributed to soil redistribution. If these assumptions hold, then any reduction in inventory at a sampling location with respect to the reference inventory can be attributed to soil loss and a gain in inventory would suggest soil deposition, given that the redistribution of ^{7}Be (as a cohesive tracer), will only occur in line with movement of the sediment to which it is adsorbed.

To quantify the mass of soil eroded or deposited (kg m^{-2}), it is essential to determine the depth distribution of ^{7}Be in the soil profile and, in particular, the relaxation mass depth (referred to as h_0) a value, which describes the shape of the ^{7}Be distribution. This is an essential component of the commonly applied soil redistribution model described below. On the basis of this, any change in ^{7}Be inventory at a sampling location can be fitted to the mass of soil redistribution required to implement that change. An example is given in Fig. 1.3, which shows a loss of 100 Bq m^{-2} at the eroded location (i.e. inventory of 300 Bq m^{-2}) with respect to the stable reference inventory (i.e. 400 Bq m^{-2}). By establishing the exponential shape of the ^{7}Be profile, the conversion model is able to estimate the mass depth of soil required to reduce the inventory by 100 Bq m^{-2}.

The specific conversion model used to derive soil redistribution magnitudes (Profile Distribution Model [PDM]) is detailed in Blake et al. (1999) and Walling et al. (2009) and the basis of this is outlined in Appendix 1.1. In addition, a worked example of this model is provided in Chap. 4.

1.3 Examples of the Application of ^{7}Be as a Soil Redistribution Tracer

Blake et al. (1999) and Walling et al. (1999) carried out the first comprehensive investigation into the use of ^{7}Be to estimate short-term soil redistribution by studying the impact of a heavy rainfall event at an agricultural site in southwest England. The ^{7}Be depth distributions from this study site are shown in Fig. 1.4. The key requirement of uniform ^{7}Be fallout across the study location was satisfied owing to the soil ^{7}Be inventory being reset (i.e. reduced to below detectable limits) by ploughing prior to

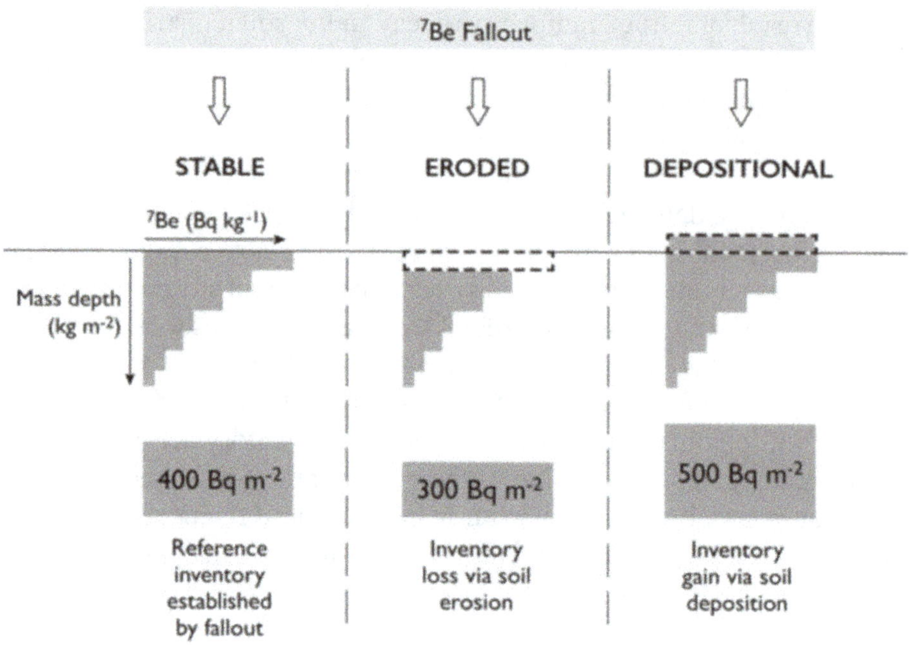

Fig. 1.3 Conceptual diagram of the premise for using ^{7}Be as a soil erosion tracer. The figure shows the ^{7}Be inventory (400 Bq m^{-2}) and depth distribution at a stable reference location, which has been established by ^{7}Be atmospheric fallout. The eroded and depositional zones (representing a neighbouring hillslope area for example) have received the same fallout as the reference location, thus, an inventory has been established but the zones have experienced depletion or enhancement of the ^{7}Be inventory as a result of soil erosion and soil deposition respectively. This simplified conceptual diagram can be compared to examples of measured ^{7}Be inventories in Fig. 1.4

the commencement of the investigation. There was also a period of low intensity rainfall prior to the erosive event, which established a measurable ^{7}Be inventory without causing significant soil redistribution. A reference baseline inventory was estimated from neighbouring undisturbed pasture locations against which the slope inventories could be compared. The depth profile was determined in a non-eroded part of the study field. The PDM was used to determine soil redistribution rates and results indicated high levels of soil export from the site as a consequence of the intense rainfall event and compacted soil conditions, highlighting the pressing need for soil conservation measures to maintain soil fertility and to prevent sediment from entering catchment watercourses.

As highlighted by Mabit et al. (2014), an important advantage in the use of ^{7}Be as a soil erosion tracer is the ability to assess the effectiveness of recent soil conservation measures or the impacts of land use change. Schuller et al. (2006) applied the approach proposed by Blake et al. (1999) to document the effects of recent forest clearance upon slope soils in a timber harvesting region of Chile. Through the ^{7}Be approach, the use of mitigation strategies could be evaluated and, in this case, the need for further improvements in soil conservation strategies was identified. Sepulveda et al. (2008) aimed to assess the effects of land management practice (in this case stubble burning) by using a combined FRN approach (^{7}Be in conjunction with

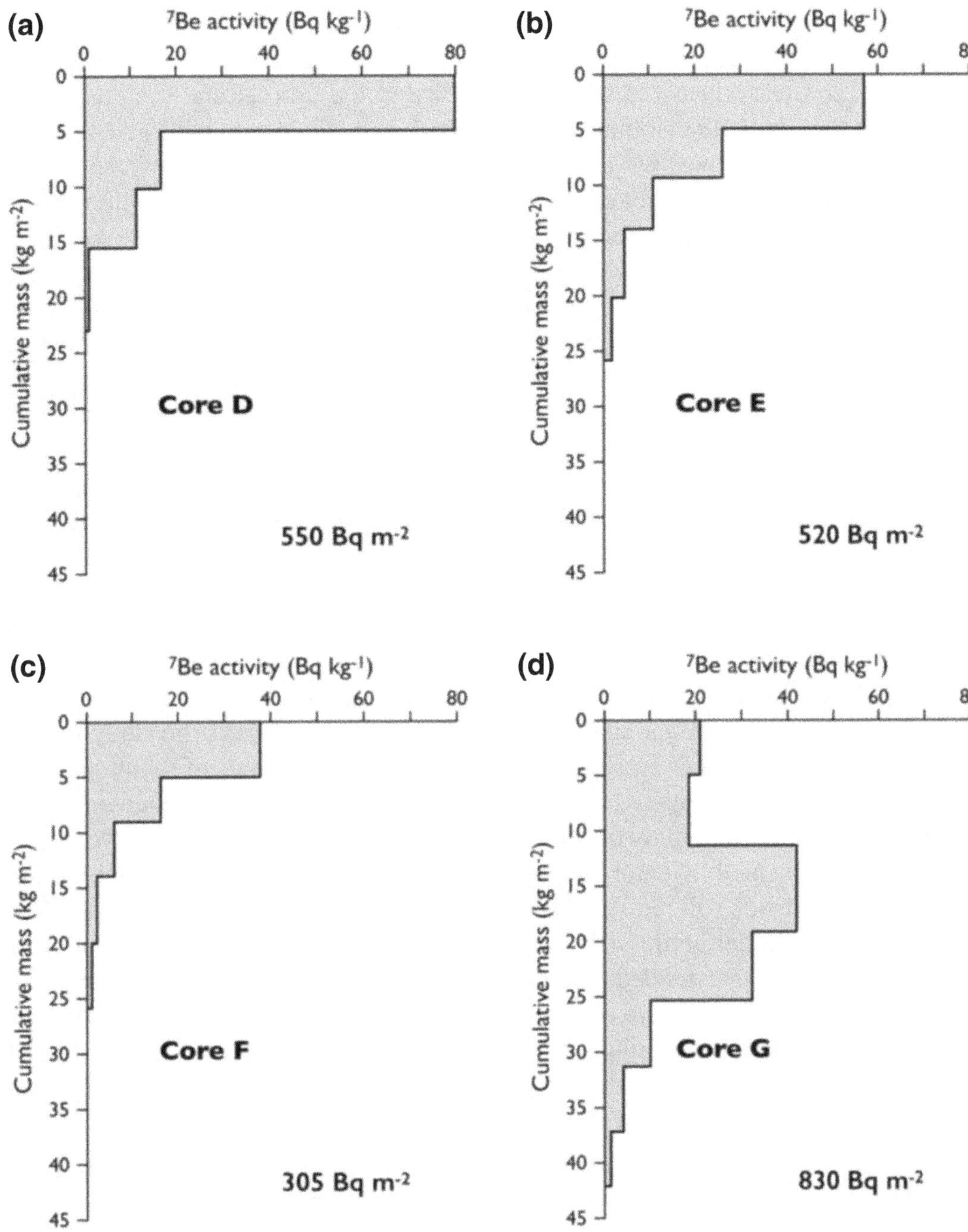

Fig. 1.4 ^{7}Be depth distributions at a stable reference location (**a**), eroded hillslope locations (**b** and **c**) and a depositional hillslope location (**d**) [based on Walling et al. (1999)]. Note the depleted ^{7}Be inventories at the eroding sites and the increased inventory at the depositional site with respect to the reference location

^{137}Cs). Here, the combined use of FRNs was able to compare impacts associated with different land management scenarios across a range of timescales, providing key information with regard to the sustainability of the agricultural system. Blake et al. (2009) also applied a combined approach and used ^{7}Be in conjunction with ^{137}Cs and ^{210}Pb to determine post wildfire sediment budgets at hillslope sites in southeast Australia. Data identified major sediment source areas and focussed attention on the implications of wildfire for water quality management.

Blake et al. (1999) and Sepulveda et al. (2008) identified the occurrence of extreme erosion events by comparing short term erosion rates with medium-term estimates derived from ^{137}Cs. Walling et al. (1999) showed that medium-term erosion rates at the study site were approximately 5 times lower than the short-term rates, suggesting that the studied rainfall event may not have been representative of seasonal patterns in rainfall. Owing to this, adjustments to the model were suggested by Walling et al. (2009) to enable the PDM to be applied across a greater timescale, encompassing a number of erosive rainfall events and improving representativeness. Thus, a method was developed for extending the use of the PDM to periods covering a few months or wet seasons. Within this extended timescale approach, the fundamental aspects of the model remain similar to the conventional approach whereby ^{7}Be inventories at a sample point are compared to a reference baseline and the depth distribution of ^{7}Be is measured. However, for the approach to be applied across multiple erosive rainfall events it is crucial to consider changes in the ^{7}Be inventory across the study period to avoid underestimation of erosion rates. Additional components of the model are, therefore, included whereby loss of inventory at a sampling point is accounted for by estimating the relative erosivity of each rainfall event and ongoing radioactive decay. Similarly, assessment of inventory gain accounts for additional ^{7}Be fallout during the study period alongside soil redistribution. The application of this modified approach was piloted by Schuller et al. (2010), with soil conservation measures shown to be less effective across an extended time period, which encompassed a range of rainfall events, demonstrating the value of considering soil erosion in a long-term context.

Recent plot studies have validated the method on a site-specific basis by demonstrating comparability between the ^{7}Be-derived soil redistribution budget and direct measurements (e.g. Porto and Walling 2014; Shi et al. 2011).

1.4 The Requirement for a Standardised Approach

FRN tracers are likely to be used in a wide range of environments to assess soil redistribution and, as such, numerous methodological approaches for establishing the key model parameters have been applied. Although it is necessary to tailor sampling approaches to local environmental conditions, it is also important that the approach is standardised, as far as is practicable, to ensure data quality and aid comparability (Mabit et al. 2008).

^{7}Be has the potential to be a valuable decision support tool for catchment managers although to achieve robust estimates of soil redistribution using the PDM it is crucial to accurately estimate its key parameters namely:(1) the reference ^{7}Be inventory, and (2) the ^{7}Be depth distribution (the latter to ensure an accurate value of h_0). It is also important to consider whether the selective transport of particle size fractions is occurring during rainfall events given preferential association of ^{7}Be to fine soil fractions (<63 μm). Failure to account for preferential adsorption to fine particles can lead to overestimation of soil erosion rates since a large proportion of the ^{7}Be inventory is associated with the fine sediment fractions, which are likely to be readily mobilised during rainfall events. Hence a relatively large deficit in ^{7}Be inventory could actually equate to a low mass of soil loss.

The requirement for standardising procedures is of particular importance for determining ^{7}Be depth distributions in light of recent studies, which demonstrate how the choice of sampling method can influence results (Baumgart et al. 2016; Ryken et al. 2016). Determining ^{7}Be depth distributions in soils is a practical challenge and one that requires careful consideration of the sampling site, both in terms of stability and uniformity (Ryken et al. 2016) as well as a precise sampling method. The latter requires a technique that allows fine depth increments (2 mm depths) to be collected throughout the ^{7}Be profile enabling enough detail to be able to accurately determine h_0. Mabit et al. (2014) designed and tested a reproducible tool and a modus operandi, which permits collection of fine soil increments in a range of soil types and provides a basis for standardising the sampling method for h_0 determination.

In practice, developing a robust sampling strategy to establish model parameters and assess soil redistribution is likely to require a balance between the number of samples required to ensure data quality and the capacity for sample analysis. As with any field operations, preliminary knowledge of the agro-environmental conditions and background data (e.g. land use history; soil hydrological properties) at the sampling site is beneficial and will help to streamline a pragmatic sampling programme.

Against this background, Chap. 2 provides a guideline for undertaking a soil redistribution study using ^{7}Be as a tracer by outlining a standard approach for establishing key conversion model parameters.

Glossary

FRN	Fallout Radionuclide.
Becquerel	The SI unit of radioactivity.
Activity Concentration	Radioactivity per unit of volume or mass e.g. Bq L^{-1}; Bq kg^{-1}.
Inventory	Radioactivity per unit area (Bq m^{-2}) also termed areal activity.
Relaxation mass depth (h_0)	The soil mass depth (kg m^{-2}) at above which 63.2% of the beryllium-7 inventory can be found.
Cumulative mass depth	The distribution of ^{7}Be with depth in the soil profile is expressed in units of areal mass

	(kg m^{-2}) rather than metric units (mm). This provides a more accurate measure of depth in the context of the soil redistribution models.
Profile Distribution Model (PDM)	Simple model to convert ^{7}Be inventory loss into soil erosion amounts by linking inventory change to the depth profile to derive the mass depth of soil loss. This model is applied to single events over a short period of time.

Appendix 1.1: The Profile Distribution Model

The exponential depth distribution of ^{7}Be in a soil profile can be described as:

$$C(x) = ce^{-x/h_0} \tag{1.1}$$

where C_x (Bq kg^{-1}) is the ^{7}Be activity at mass depth x (kg m^{-2}), c is a constant value and h_0 is the relaxation mass depth, the depth at above which 63.2% of the ^{7}Be inventory can be found.

Erosion rates (kg m^{-2}) can be estimated by comparing the ^{7}Be inventories at the sample site, A (Bq m^{-2}), to the reference inventory, A_{ref} (Bq m^{-2}). Where a mass of soil has been lost (h) (kg m^{-2}) changes in the sample site inventories can be represented as:

$$A = \int_{-h}^{\infty} C(x)dx = A_{ref}e^{h/h_0} \tag{1.2}$$

Erosion rate, h (kg m^{-2}, negative), can, therefore, be calculated as:

$$h = h_0 \ln(A/A_{ref}) \tag{1.3}$$

Deposition of material is reflected in an excess of ^{7}Be inventory at the sample site with respect to the reference site. The depth of deposition, h' (kg m^{-2}, positive), can be calculated as:

$$h' = \left(A - A_{ref}\right)/C_d \tag{1.4}$$

Where C_d (Bq kg^{-1}) is the mean activity concentration of ^{7}Be in the deposited sediment.

The ^{7}Be activity concentration in the eroding sediment at each upslope point, C_e (Bq kg^{-1}), can be calculated from the loss of inventory divided by the mass of soil loss:

$$C_e = \left(A_{ref} - A\right)/h \tag{1.5}$$

The mean activity concentration of soil mobilised from the study area, S (m^2), can then be calculated as:

$$C_{\mathrm{d}} = \int_S hC_{\mathrm{e}} dS / \int_S hdS \tag{1.6}$$

References

Baumgart, P., Riedel, E., Eltner, A., & Faust, D. (2016). Soil surface sampling approaches for reliable radiogenic isotope tracer investigations. *Soil Science, 181,* 82–88. https://doi.org/10.1097/SS.0000000000000141.

Blake, W. H., Walling, D. E., & He, Q. (1999). Fallout beryllium-7 as a tracer in soil erosion investigations. *Applied Radiation and Isotopes, 51,* 599–605.

Blake, W. H., Wallbrink, P. J., Wilkinson, S. N., Humphreys, G. S., Doerr, S. H., Shakesby, R. A., et al. (2009). Deriving hillslope sediment budgets in wildfire-affected forests using fallout radionuclide tracers. *Geomorphology, 104,* 105–116.

Doering, C., & Akber, R. (2008). Beryllium-7 in near-surface air and deposition at Brisbane, Australia. *Journal of Environmental Radioactivity, 99,* 461–467. https://doi.org/10.1016/j.jenvrad.2007.08.017.

Ioannidou, A., & Papastefanou, C. (2006). Precipitation scavenging of Be-7 and Cs radionuclides in air. *Journal of Environmental Radioactivity, 85,* 121–136.

Iurian, A.-R., Toloza, A., Adu-Gyamfi, J., & Cosma, C. (2013). Spatial distribution of ^{7}Be in soils of Lower Austria after heavy rains. *Journal of Radioanalytical and Nuclear Chemistry, 298,* 1857–1863. https://doi.org/10.1007/s10967-013-2683-8.

Kaste, J. M., Norton, S. A., & Hess, C. T. (2002). Environmental chemistry of beryllium-7. *Beryllium: Mineralogy, petrology, and geochemistry* (pp. 271–289). Washington: Mineralogical Soc America.

Kusmierczyk-Michulec, J., Gheddou, A., & Nikkinen, M. (2015). Influence of precipitation on ^{7}Be concentrations in air as measured by CTBTO global monitoring system. *Journal of Environmental Radioactivity, 144,* 140–151. https://doi.org/10.1016/j.jenvrad.2015.03.014.

Mabit, L., Benmansour, M., & Walling, D. E. (2008). Comparative advantages and limitations of the fallout radionuclides ^{137}Cs, ^{210}Pb$_{ex}$ and ^{7}Be for assessing soil erosion and sedimentation. *Journal of Environmental Radioactivity, 99,* 1799–1807.

Mabit, L., Meusburger, K., Iurian, A.-R., Owens, P. N., Toloza, A., & Alewell, C. (2014). Sampling soil and sediment depth profiles at a fine resolution with a new device for determining physical, chemical and biological properties: The Fine Increment Soil Collector (FISC). *Journal of Soils and Sediments, 14,* 630–636. https://doi.org/10.1007/s11368-013-0834-8.

Porto, P., & Walling, D. E. (2014). Use of ^{7}Be measurements to estimate rates of soil loss from cultivated land: Testing a new approach applicable to individual storm events occurring during an extended period. *Water Resources Research, 50,* 8300–8313. https://doi.org/10.1002/2014WR015867.

Ryken, N., Al-Barri, B., Taylor, A., Blake, W., Maenhout, P., Sleutel, S., et al. (2016). Quantifying the spatial variation of ^{7}Be depth distributions towards improved erosion rate estimations. *Geoderma, 269,* 10–18. https://doi.org/10.1016/j.geoderma.2016.01.032.

Schuller, P., Iroume, A., Walling, D. E., Mancilla, H. B., Castillo, A., & Trumper, R. E. (2006). Use of beryllium-7 to document soil redistribution following forest harvest operations. *Journal of Environmental Quality, 35,* 1756–1763. https://doi.org/10.2134/jeq2005.0410.

Schuller, P., Walling, D. E., Iroume, A., Castillo, A. (2010). Use of beryllium-7 to study the effectiveness of woody trash barriers in reducing sediment delivery to streams after forest clearcutting. *Soil and Tillage Research, 110*, 143–153.

Sepulveda, A., Schuller, P., Walling, D. E., & Castillo, A. (2008). Use of Be-7 to document soil erosion associated with a short period of extreme rainfall. *Journal of Environmental Radioactivity, 99,* 35–49. https://doi.org/10.1016/j.jenvrad.2007.06.010.

Shi, Z. L., Wen, A. B., Zhang, X. B., & Yan, D. C. (2011). Comparison of the soil losses from ^{7}Be measurements and the monitoring data by erosion pins and runoff plots in the Three Gorges Reservoir region, China. *Applied Radiation and Isotopes, 69,* 1343–1348. https://doi.org/10.1016/j.apradiso.2011.05.031.

Taylor, A., Blake, W. H., Couldrick, L., & Keith-Roach, M. J. (2012). Sorption behaviour of beryllium-7 and implications for its use as a sediment tracer. *Geoderma, 187–188,* 16–23.

Taylor, A., Blake, W. H., Smith, H. G., Mabit, L., & Keith-Roach, M. J. (2013). Assumptions and challenges in the use of fallout beryllium-7 as a soil and sediment tracer in river basins. *Earth-Science Reviews, 126,* 85–95.

Taylor, A., Blake, W. H., & Keith-Roach, M. J. (2014). Estimating Be-7 association with soil particle size fractions for erosion and deposition modelling. *Journal of Soils and Sediments, 14,* 1886–1893. https://doi.org/10.1007/s11368-014-0955-8.

Taylor, A., Keith-Roach, M. J., Iurian, A. R., Mabit, L., & Blake, W. H. (2016). Temporal variability of beryllium-7 fallout in southwest UK. *Journal of Environmental Radioactivity, 160,* 80–86. https://doi.org/10.1016/j.jenvrad.2016.04.025.

Walling, D. E., He, Q., & Blake, W. (1999). Use of Be-7 and Cs-137 measurements to document short- and medium-term rates of water-induced soil erosion on agricultural land. *Water Resources Research, 35,* 3865–3874.

Walling, D. E., Schuller, P., Zhang, Y., & Iroume, A., (2009). Extending the timescale for using beryllium-7 measurements to document soil redistribution by erosion. *Water Resources Research 45*. W02418 https://doi.org/10.1029/2008wr007143.

The opinions expressed in this chapter are those of the author(s) and do not necessarily reflect the views of the International Atomic Energy Agency, its Board of Directors, or the countries they represent.

Open Access This chapter is licensed under the terms of the Creative Commons Attribution 3.0 IGO License (https://creativecommons.org/licenses/by/3.0/igo/), which permits use, sharing, adaptation, distribution and reproduction in any medium or format, as long as you give appropriate credit to the International Atomic Energy Agency, provide a link to the Creative Commons licence and indicate if changes were made.

The use of the International Atomic Energy Agency's name, and the use of the International Atomic Energy Agency's logo, shall be subject to a separate written licence agreement between the International Atomic Energy Agency and the user and is not authorized as part of this CC-IGO licence. Note that the link provided above includes additional terms and conditions of the licence.

The images or other third party material in this chapter are included in the chapter's Creative Commons licence, unless indicated otherwise in a credit line to the material. If material is not included in the chapter's Creative Commons licence and your intended use is not permitted by statutory regulation or exceeds the permitted use, you will need to obtain permission directly from the copyright holder.

Chapter 2
How to Design a Be-7 Based Soil Distribution Study at the Field Scale: A Step-by-Step Approach

W. H. Blake, A. Taylor, A. Toloza and L. Mabit

2.1 Key Sample Sets and Associated Data

A ^{7}Be-based soil redistribution budget is based on several key datasets (Table 2.1) which have strict rules on collection locations and timings, depending on the time period of application. Overall, the methodological approach follows the principles of other FRN techniques (e.g. the ^{137}Cs approach) but with necessary differences linked to the short half-life of ^{7}Be and its delivery dynamics. The difference in delivery dynamics also provides the added advantage of opportunity for assumptions underpinning the approach to be tested in field and by laboratory experimentation (Taylor et al. 2014). As summarized in Table 2.1, some datasets are mandatory to convert measurement of ^{7}Be inventory into soil redistribution amounts. Other datasets are advised under some circumstances to assist with data interpretation and improve the quality of soil distribution estimates. Fundamental considerations for the collection of all these datasets are outlined in the following section.

W. H. Blake (✉)
School of Geography, Earth and Environmental Sciences, University of Plymouth, Plymouth, UK
e-mail: william.blake@plymouth.ac.uk

A. Taylor
Consolidated Radioisotope Facility (CoRiF), University of Plymouth, Plymouth, UK
e-mail: alex.taylor@plymouth.ac.uk

A. Toloza · L. Mabit
Soil and Water Management and Crop Nutrition Subprogramme, Joint FAO/IAEA Division of Nuclear Techniques in Food and Agriculture, International Atomic Energy Agency, Vienna, Austria
e-mail: A.Toloza@iaea.org

L. Mabit
e-mail: L.Mabit@iaea.org

© International Atomic Energy Agency (IAEA) 2019

L. Mabit and W. Blake (eds.), *Assessing Recent Soil Erosion Rates through the Use of Beryllium-7 (Be-7)*, https://doi.org/10.1007/978-3-030-10982-0_2

Table 2.1 Samples, measurements and resulting datasets required to construct a ^{7}Be-based soil redistribution budget at the hillslope scale

Sampling location	Samples or measurements taken	Associated dataset	Method(s)	Priority status
Reference site	Reference soil cores	^{7}Be reference inventory for undisturbed soil condition	Short manual core tubes (ca. 30–50 mm)	Needed for all applications
	Sectioned soil core	Depth profile of ^{7}Be and h_0	Fine increment soil sampler (ca 2 mm)	Needed for all applications
Study site	Study plot soil cores	^{7}Be inventory of eroding and depositing sites; particle size distribution	Short manual core tubes (ca. 30–50 mm)	Needed for all applications
Reference site	Rainfall samples (time-integrated)	^{7}Be activity concentration in rainfall	Large funnel and polypropylene container	Recommended for short-term event-based study; essential for extended time series model
	Rainfall record	15 min rainfall data for amount and intensity	Tipping bucket rain gauge	Recommended for short-term event-based study; essential for extended time series model
Reference site and study site	Soil infiltration capacity	Infiltration capacity	Minidisc infiltrometer or similar	Recommended for all applications
Study site	Representative sample of mobilised and deposited soil	Particle size properties	Runoff and sediment traps; rainfall simulation	Recommended for all applications where particle size selectivity of erosion is likely to occur
Reference site and study site	Topographic survey	Digital Elevation Model (DEM) of study plot	GPS-based or traditional survey	Recommended for all applications

2.2 Reference Site Selection and Sampling

As outlined in Chap. 1, ^{7}Be inventory data are used to construct an FRN budget for the hillslope, to evaluate relative differences to the reference inventory, and this budget is subsequently converted into a soil redistribution budget (described in detail in Chap. 4). Accurate and representative determination of the reference inventory, and associated uncertainty is therefore a fundamental requirement since estimates of soil redistribution pattern and amount relies on this key value.

Reference sites serve two purposes in ^{7}Be studies: (1) to determine a mean reference inventory (i.e. the areal activity of ^{7}Be (Bq m^{-2}) in the soil surface unaffected by erosion), and (2) to determine the depth distribution of ^{7}Be in the undisturbed soil prior to erosion, to derive h_0. The first requirement is based on collection of bulk soil cores, and the second on the collection of at least one sectioned core. Both need to be determined at a flat, stable location near the study plot, as for ^{137}Cs studies, but it is essential that the land-use history of the location for the depth profile is exactly the same as the study plot. This is important as the h_0 value determined must be representative of the eroding soil surface. In practice, the best location for both these measures is a flat area that has been cultivated at the same time and in the same manner as the study plot (Fig. 2.1).

Sample designs for bulk cores need to account for potential FRN spatial variability within the reference site (Sutherland 1996; Mabit et al. 2012; Kirchner 2013) and also deliver sufficient mass of soil for sample analysis (Chap. 3). However, sample numbers are often constrained in ^{7}Be tracing investigations owing to the short half-life

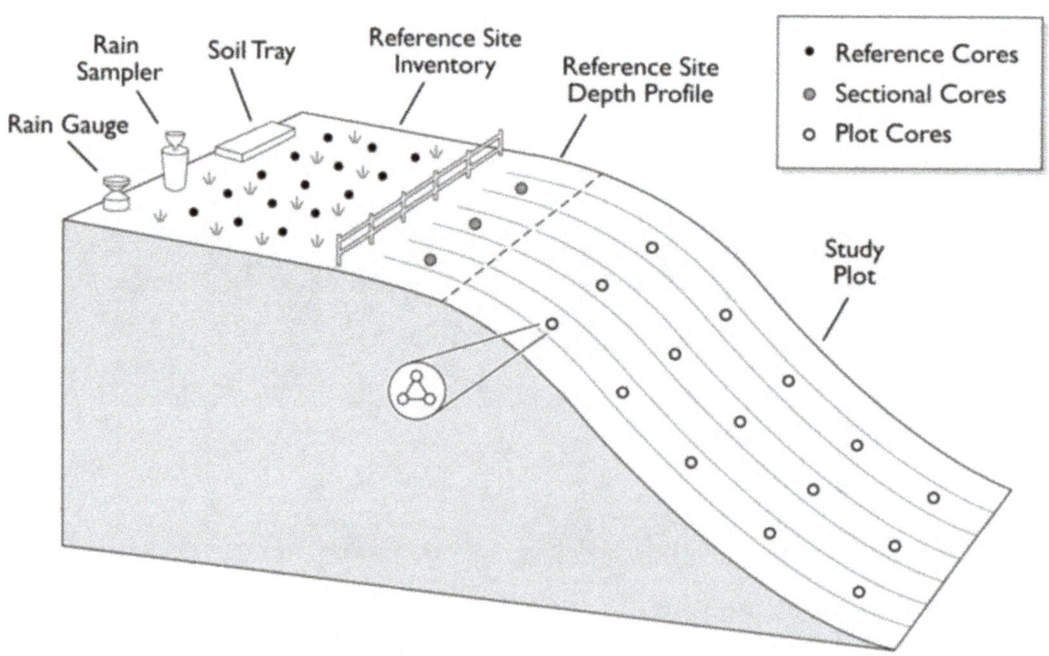

Fig. 2.1 Experimental design to establish a hillslope ^{7}Be budget over largely bare soil

and availability of sufficient gamma detection facilities. Considering this limitation, Taylor et al. (2013) recommend that all studies should state a reference inventory with suitable upper and lower limits (i.e. $\pm 2\sigma$) (Owens and Walling 1996) and this should be incorporated into subsequent soil redistribution estimates (Chap. 4). Spatial variability in inventory within a plot is most likely due to local redistribution of soil by rain splash and micro-topography and accumulation of rain-splashed particles in hollows. To capture such variability at any given reference site, it is recommended that each sample integrates several cores and then this process is replicated to provide a minimum of 10–15 spatially-integrated reference samples for analysis (Fig. 2.1). Bulk soil cores need to be taken to a consistent depth that is below the known depth penetration of ^{7}Be in the study soil. Depth penetration is typically 20–30 mm (Blake et al. 1999; Doering et al. 2006; Sepulveda et al. 2008; Wallbrink and Murray 1996). In this regard, we strongly advise undertaking a preliminary investigation for trial depth-profile dataset at the study area to ensure the complete profile is captured whilst avoiding dilution of the sample by overestimating the profile depth. This will allow the researcher (1) to establish the maximum depth penetration of ^{7}Be in the soil of the investigated area and (2) to use this depth to determine the maximum depth collection of the remaining bulk cores to be collected (e.g. 0–30 mm in the reference site). Any vegetation on the ground surface must be included in the sample as this will carry part of the recent ^{7}Be inventory (Iurian et al. 2015). When collecting spatially-integrated cores, it is important that the total sum of the core areas that comprise one sample is recorded and that the depth penetration of all cores is consistent at 1 mm precision.

When characterising the ^{7}Be depth profile, it is essential that section cores are sampled from a soil surface that has experienced the same cultivation practice as the study slope but in a location that has remained undisturbed by erosion or deposition processes. Variability due to rain splash, as noted above, can present challenges in selection of the appropriate position of the core. It is highly recommended that more than one section core is collected but this is often limited by gamma detector resource availability. An alternative is to collect 3 replicate section cores and combine the respective layers from each core into integrated samples to capture, but not to quantify, spatial variability within the flat, uneroded reference area. The recommended tool for standardized section core sample collection is the Fine Increment Soil Collector (FISC) (Mabit et al. 2014; Fig. 2.2) which is proven to collect high precision depth profile data suitable for supporting ^{7}Be inventory conversion (Ryken et al. 2016).

It is also common practice to establish a rainfall collection and monitoring station at the reference site to permit assessment of the dynamic of ^{7}Be delivery in relation to inventory development and its radioactive decay in the study area under investigation (Fig. 2.3). Sampling of reference cores through a time period also serves to benchmark and/or validate rainfall-based inventory assessment (Wallbrink and Murray 1994; Walling et al. 2009). The importance of this when applying the event scale PDM is to validate the stability of the reference site. With high resolution rainfall data and rainwater samples, we can use the modelled inventory as a benchmark for confidence in our choice of reference site. For this purpose, rainfall data need to be collected using a tipping bucket rain gauge (Fig. 2.4) that provides 15 or 30 min interval

Fig. 2.2 Fine Increment Soil Collector (Mabit et al. 2014)

rainfall intensity data. Alongside rainfall monitoring, bulk rainfall samples need to be collected at a minimum of monthly intervals but preferably at event-scale intervals (depending on rainfall regime) and analysed for ^{7}Be concentration (Bq L^{-1}). Samples should be collected in pre-acidified polypropylene containers with an attached funnel for rainfall capture (Fig. 2.3) and ^{7}Be extracted following protocols described by (Taylor et al. 2016); (see Appendix 2.1 for standard operating procedure).

A time series of the relative inventory at a study site (e.g. Fig. 2.3) can then be calculated as follows, where for each *daily time step*:

$$A(t) = A(t-1) \cdot exp(-\lambda) + F(t) \tag{2.1}$$

where $A(t-1)$ is the ^{7}Be inventory of the previous day (Bq m^{-2}), λ is the radioactive decay constant (daily time step) and $F(t)$ is the fallout contribution of the current day. It is common practice to determine the start point either through collection of a suite of reference inventory cores (Walling et al. 2009) or to assume zero due to tillage of the soil surface and dilution of the ^{7}Be signal within the soil profile. Table 2.2 shows a spreadsheet coding example to create a cumulative inventory dataset.

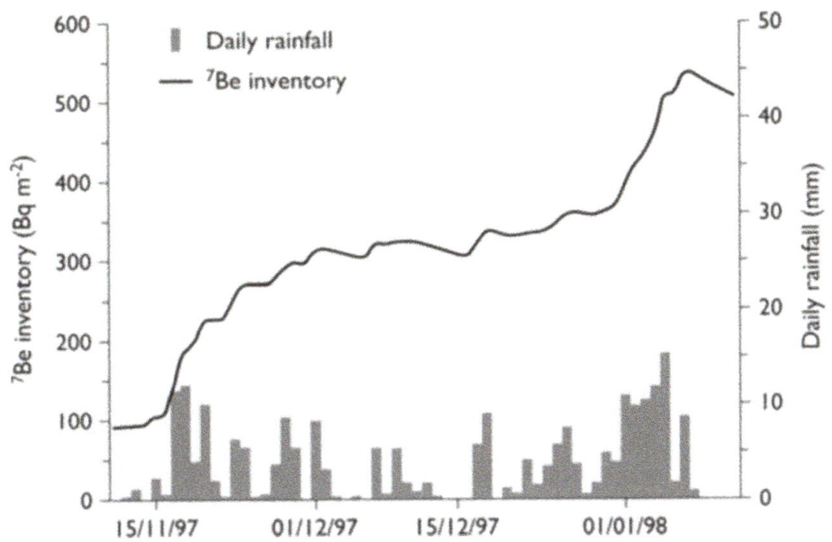

Fig. 2.3 Example of reference inventory development and decay record based on rainfall sampling and analysis for ^{7}Be alongside rainfall monitoring (Blake et al. 1999)

Fig. 2.4 Examples of tipping bucket rain gauge (left) and rainfall sampler (right)

2.3 Sample Design Options for Soil Redistribution

The ^{7}Be approach is most suited to erosion studies performed on bare soil surfaces (no vegetation) or with little vegetation cover. The method is limited to quantification of erosion by rain splash, sheet wash and shallow rill development since once rill incision goes beyond the depth of the ^{7}Be depth profile, eroded soil is exported in the absence of the tracer signal. The selection of study site is very much dependent on the

Table 2.2 Coding for Microsoft Excel spreadsheet to calculate cumulative inventory from rainfall data and rainwater ^{7}Be activity concentration data

Cell number (assuming first row is column headings NB there is only one entry on row 2—see below)	Code	Explanation
A3	n/a	Date and time step of rainfall data
B3	n/a	Rainfall input for time step (mm)
C3	=B3 * 1000/1000	Rainfall expressed as volume per unit area (L m^{-2})
D3	n/a	Rainfall ^{7}Be activity concentration (Bq L^{-1})
E3	D3 * C3	^{7}Be areal activity deposition for time step (Bq m^{-2})
F2	n/a	Inventory at start of monitoring (generally zero if plot is cultivated at beginning)
F3	=E3 + (F2 * EXP(-$I$1))	Plot inventory (Bq m^{-2}) at timestep where cell I1 contains the decay constant for the appropriate time step. The ^{7}Be deposition received (cell E3) is added to the decay corrected inventory from the previous day

research question but, in any case, a basic pilot study will serve useful in determining the presence of a sufficient ^{7}Be inventory in the region to make the approach viable and, as described above, to evaluate the typical depth penetration of ^{7}Be to inform bulk core sample depth.

The ^{7}Be approach may be applied at different spatial scales depending upon the questions being asked by land managers and the constraints imposed by the key assumptions discussed in Chap. 1. In this context, there are three main ways to design a soil core sampling strategy within the study hillslope: (1) sampling along transects from upper to lower slope (Schuller et al. 2006); (2) high resolution grid sampling (Walling et al. 1999; Blake et al. 1999); (3) spatially-integrated sampling within defined geomorphic landscape units (Wallbrink et al. 2002; Blake et al. 2009).

At the field scale, the single transect (Fig. 2.1) or a multi-transect sampling is the most straightforward and cost-effective approach in terms of field sample collection and laboratory processing work effort. Transects can, however, be limited in terms of spatial representativeness depending on local topography and research questions to be addressed. A high-resolution regular grid approach is effective for evaluation of spatial variability and will provide more representative information on soil export and sediment delivery ratio from a larger study area than transects. However, this approach is highly limited by the analytical demand of a large number of samples, which is hampered further by the short half-life of ^{7}Be. The geomorphic landscape unit approach offers a compromise between the limited spatial representativeness of the transect approach and the analytical demands of the grid approach. However, it should be pointed out that multiple reference sites may be required in some geomorphic landscapes.

2.4 Sampling for Particle Size Selectivity Correction

Particle size selectivity during soil erosion processes is well-known (Bernard et al. 1992) and users of soil loss assessment or measurement approaches need to make a decision on the relevance of this specific process to their own study site. Taylor et al. (2014) describe how particle size selectively of soil erosion processes can be accounted for in the ^{7}Be conversion model. Application of this method requires specific samples to be collected during or after the erosion event being studied.

Particle size correction requires representative samples of (1) uneroded soil, (2) mobilised soil and (3) deposited soil to be collected (Fig. 2.5). The first can simply be represented by the bulk cores collected for depth profile determination (i.e. in a non-eroding but cultivated site). The third requirement above can be represented by the samples collected to determine inventory in areas of sediment accumulation. The second, the characterisation of mobilised soil, requires more careful planning since after an erosion event, such material has either left the study site or been, in part, deposited at the foot slope. Taylor et al. (2014) propose that such material can be collected with Gerlach troughs installed in the study site or simple equivalent sediment trap systems. Alternatively, to avoid the need to install equipment prior to the event, a rainfall simulator could be used to mobilise material from a representative area, and the material captured for analysis. This needs to be done under rainfall conditions similar to the erosive event.

All samples collected need to be analysed for particle size distribution (Chap. 4).

2.5 Summary: Designing a Basic Small Scale ^{7}Be Pilot Study Sampling Programme at the Plot Scale

The procedures described in this chapter can be practised and developed within the context of a simple pilot study framework designed to test the viability of using ^{7}Be as a tracer in any given landscape. A step-by-step approach is detailed below:

Step 1: Having located your study site and secured permission from the landowners for its investigation, collect 3 topsoil samples (20 mm depth) after significant rainfall (e.g. >sufficient rainfall to develop a measurable inventory for the study area over an extended time period) and analyse them for ^{7}Be content to confirm if significant and measurable inventory of this radioisotope is present in the study area;

Step 2: Collect a trial section core to evaluate the depth penetration of ^{7}Be in the study soil;

Step 3: Select the reference site in the study area and, if possible, the study plot location (prior to soil erosion taking place). Set up a rain gauge and, if required, rain sampling equipment. Note the timing of the last cultivation and, if this was some time prior to current time within which rainfall had occurred, collect a set of reference cores to establish the baseline (t = 0) inventory. Note this will be zero by default if plot study begins immediately after cultivation and mixing of existing ^{7}Be inventory into the soil profile;

Step 4: After erosion has taken place within a target study plot (a) collect 10–15 spatially-integrated bulk reference cores from the non-eroding reference site to the depth of maximum ^{7}Be penetration (e.g. 20–30 mm) following the information provided by the test core (in step 2), (b) collect up to 3 section cores from a flat, non-eroding area that has experienced the same cultivation practice as the study plot using the FISC and either analyse cores separately or combine layers to create a spatially integrated depth profile depending on analytical resource, (c) collect spatially-integrated cores along three transects within the study plot (Fig. 2.1) with 5–15 samples per transect depending on your analytical resource;

Step 5: Undertake necessary sampling for particle size correction if desired;

Step 6: Bring all recovered soil samples to laboratory (i) for preparation prior to gamma spectrometry (Chap. 3) measurements and, when necessary, (ii) for performing particle size distribution analysis by laser granulometry.

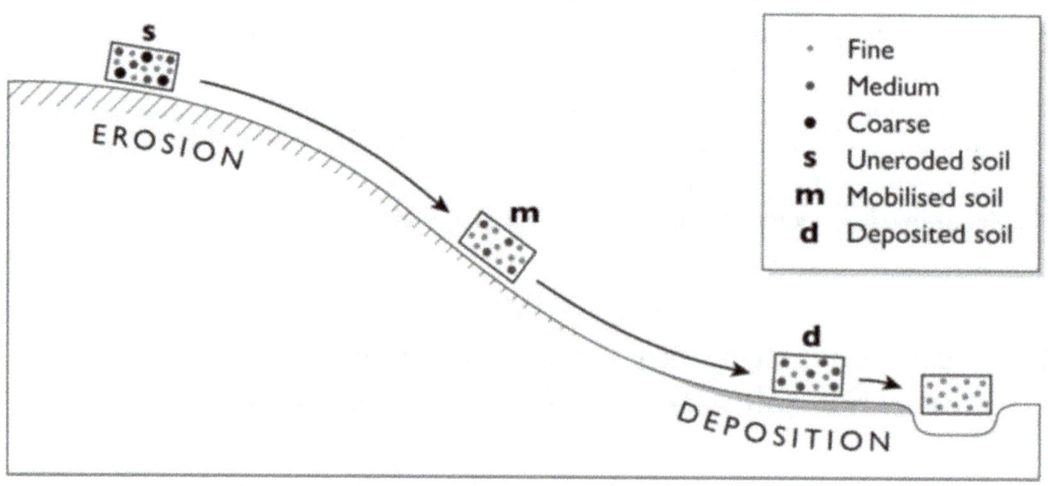

Fig. 2.5 Schematic diagram of sampling protocol for including particle size selectivity in ^{7}Be soil erosion study (uneroded soil = **s**; mobilised soil = **m**; deposited soil = **d**)

Glossary

Reference Inventory	Radioactivity per unit area (Bq m^{-2}), also termed areal activity, at a stable (non-eroding) field site in close proximity to the eroding study site.
Bulk soil core	A soil core taken to a specific depth wherein the material recovered within the corer represents one sample.
Section core	A soil core that is initially kept intact within the soil corer and then extruded in small (ca 2 mm) increments to permit sections to be subsampled layer by layer.
Spatially–integrated sampling	A process through which one sample collected for analysis comprises several smaller samples taken over a wider area to capture local variability e.g. due to micro-topography or variable vegetation cover. Can be used in routine core sampling but also extended to underpin the landscape-unit sampling approach to ^{7}Be budgeting (see text for details).
Particle size selectivity	The process by which erosion processes preferentially remove fine-grained soil particles due to greater critical shear stress required for mobilisation of coarser particles.

Appendix 2.1: Protocol for Extraction of ^{7}Be from Rainwater

According to Taylor et al. (2016), prior to deployment of rainfall collectors, 10 mL HCl (2.5 M) is added to each bottle to prevent adsorption of ^{7}Be to vessel walls during the sampling period. Due care must be taken when handling acid in accordance with your institutions Health and Safety code. Collectors may be exposed for periods of between 3 and 35 days depending on the frequency and magnitude of rainfall. Generally, samples comprise fallout from a number of events across the sample period and are therefore, referred to as integrated samples. At the point of sampling, funnels should be rinsed with a known volume of HCl (1 M) and the bottles replaced with acid-cleaned bottles.

Each sample should be checked to ensure pH is < 2 and then filtered to remove any coarse debris (e.g. using Whatman grade number 41 filter papers). ^{7}Be can then be pre-concentrated from solution by co-precipitation with MnO_2 following the method detailed by Short et al. (2007).

1 mL of 0.2 M $KMnO_4$ is added per litre of rainwater sample and the pH adjusted to 8–10 using concentrated NH_4OH.

Once at the desired pH, 1 mL of 0.3 M $MnCl_2$ is added to the sample whilst stirring. MnO_2 precipitate is then allowed to settle for 24 h prior to removal by vacuum filtration using 0.45 μm cellulose nitrate filter paper.

Filter paper is then air dried, fixed with cellophane and sealed in a suitable container prior to analysis by gamma spectrometry. For each rainwater sample, duplicate 1 L subsamples should ideally be treated and the precipitate combined for filtration. Where rainfall samples are of low volume, a single 1 L sample may be treated. Samples should be considered to provide a representative sub-sample of rainfall for the period.

Repeatability can be tested by analysing triplicate subsamples separately and Relative Standard Deviation (RSD) between triplicates determined. This should typically be <10% (Taylor et al. 2016).

^{7}Be recovery from solution using the coprecipitation method was tested by Taylor et al. (2016) by reprecipitating the filtrate from 3 samples. In each case, the ^{7}Be activity in the filtrate was below Minimum Detectable Activity (MDA). MDA values for these samples were <10% of the total activity, suggesting that ^{7}Be recovery using the above method is >90%, in agreement with Short et al. (2007).

References

Bernard, C., Laverdière, M. R., & Pesant, A. R. (1992). Variabilité de la relation entre les pertes de césium et de sol par érosion hydrique. *Geoderma, 52*(3–4), 265–277. https://doi.org/10.1016/0016-7061(92)90041-5.

Blake, W. H., Walling, D. E., & He, Q. (1999). Fallout beryllium-7 as a tracer in soil erosion investigations. *Applied Radiation and Isotopes, 51*(5), 599–605.

Blake, W. H., et al. (2009). Deriving hillslope sediment budgets in wildfire-affected forests using fallout radionuclide tracers. *Geomorphology, 104,* 105–116.

Doering, C., Akber, R., & Heijnis, H. (2006). Vertical distributions of Pb-210 excess, Be-7 and Cs-137 in selected grass covered soils in Southeast Queensland, Australia. *Journal of Environmental Radioactivity, 87*(2), 135–147. https://doi.org/10.1016/j.jenvrad.2005.11.005.

Kirchner, G. (2013). Establishing reference inventories of ^{137}Cs for soil erosion studies: Methodological aspects. *Geoderma, 211–212*, 107–115.

Iurian, A.-R. et al. (2015). The interception and wash-off fraction of ^{7}Be by bean plants in the context of its use as a soil radiotracer. *Journal of Radioanalytical and Nuclear Chemistry*. https://doi.org/10.1007/s10967-015-3948-1.

Mabit, L., Chhem-Kieth, S., Toloza, A., Vanwalleghem, T., Bernard, C., Amate J. I., González de Molina, M., Gómez, J. A. (2012). Radioisotopics and physicochemical background indicators to assess soil degradation affecting olive orchards in southern Spain.*Agriculture, Ecosystems & Environment*, *159*, 70–80.

Mabit, L., et al. (2014). Sampling soil and sediment depth profiles at a fine resolution with a new device for determining physical, chemical and biological properties: The Fine Increment Soil Collector (FISC). *Journal of Soils and Sediments, 14*(3), 630–636. https://doi.org/10.1007/s11368-013-0834-8.

Owens, P. N., & Walling, D. E. (1996). Spatial variability of caesium-137 inventories at reference sites: An example from two contrasting sites in England and zimbabwe. *Applied Radiation and Isotopes, 47*(7), 699–707.

Ryken, N., et al. (2016). Quantifying the spatial variation of ^{7}Be depth distributions towards improved erosion rate estimations. *Geoderma, 269,* 10–18. https://doi.org/10.1016/j.geoderma.2016.01.032. Elsevier Science BV, PO Box 211, 1000 AE Amsterdam.

Schuller, P., et al. (2006). Use of beryllium-7 to document soil redistribution following forest harvest operations. *Journal of Environmental Quality, 35*(5), 1756–1763. https://doi.org/10.2134/jeq2005.0410.

Sepulveda, A., et al. (2008). Use of Be-7 to document soil erosion associated with a short period of extreme rainfall. *Journal of Environmental Radioactivity, 99*(1), 35–49. https://doi.org/10.1016/j.jenvrad.2007.06.010.

Short, D. B., Appleby, P. G., & Hilton, J. (2007). Measurement of atmospheric fluxes of radionuclides at a UK site using both direct (rain) and indirect (soils) methods. *International Journal of Environment and Pollution, 29*(4), 392–404.

Sutherland, R. A. (1996). Caesium-137 soil sampling and inventory variability in reference locations: A literature survey. *Hydrological Processes, 10*(1), 43–53.

Taylor, A. et al. (2013). Assumptions and challenges in the use of fallout beryllium-7 as a soil and sediment tracer in river basins. *Earth-Science Reviews, 126,* 85–95. Available at: http://www.sciencedirect.com/science/article/pii/S0012825213001281.

Taylor, A., Blake, W. H., & Keith-Roach, M. J. (2014). Estimating Be-7 association with soil particle size fractions for erosion and deposition modelling. *Journal of Soils and Sediments, 14*(11), 1886–1893. https://doi.org/10.1007/s11368-014-0955-8.

Taylor, A., et al. (2016). Temporal variability of beryllium-7 fallout in southwest UK. *Journal of Environmental Radioactivity, 160,* 80–86. https://doi.org/10.1016/j.jenvrad.2016.04.025.

Wallbrink, P. J., & Murray, A. S. (1994). Fallout of Be-7 in south-eastern Australia. *Journal of Environmental Radioactivity, 25*(3), 213–228.

Wallbrink, P. J., & Murray, A. S. (1996). Distribution and variability of Be-7 in soils under different surface cover conditions and its potential for describing soil redistribution processes. *Water Resources Research, 32*(2), 467–476.

Wallbrink, P. J., Walling, D. E., & He, Q. (2002). Radionuclide measurement using HPGe gamma spectrometry. In F. Zapata (Ed.), *Handbook for the assessment of soil erosion and sedimentation using environmental radionuclides* (pp. 67–96). Dordrecht: Kluwer.

Walling, D. E. et al. (2009). Extending the timescale for using beryllium 7 measurements to document soil redistribution by erosion. *Water Resources Research, 45*. W02418 https://doi.org/10.1029/2008wr007143.

Walling, D. E., He, Q., & Blake, W. (1999). Use of Be-7 and Cs-137 measurements to document short- and medium-term rates of water-induced soil erosion on agricultural land. *Water Resources Research, 35*(12), 3865–3874.

The opinions expressed in this chapter are those of the author(s) and do not necessarily reflect the views of the International Atomic Energy Agency, its Board of Directors, or the countries they represent.

Open Access This chapter is licensed under the terms of the Creative Commons Attribution 3.0 IGO License (https://creativecommons.org/licenses/by/3.0/igo/), which permits use, sharing, adaptation, distribution and reproduction in any medium or format, as long as you give appropriate credit to the International Atomic Energy Agency, provide a link to the Creative Commons licence and indicate if changes were made.

The use of the International Atomic Energy Agency's name, and the use of the International Atomic Energy Agency's logo, shall be subject to a separate written licence agreement between the International Atomic Energy Agency and the user and is not authorized as part of this CC-IGO licence. Note that the link provided above includes additional terms and conditions of the licence.

The images or other third party material in this chapter are included in the chapter's Creative Commons licence, unless indicated otherwise in a credit line to the material. If material is not included in the chapter's Creative Commons licence and your intended use is not permitted by statutory regulation or exceeds the permitted use, you will need to obtain permission directly from the copyright holder.

Chapter 3
Measurement of Be-7 in Environmental Materials

A. R. Iurian and G. E. Millward

3.1 Overview of Digital Gamma Spectrometry Systems

Gamma-ray spectrometry is the only 'routine' method for measuring the natural cosmogenic radionuclide ^{7}Be. Activity concentrations (Bq kg^{-1}) of ^{7}Be are determined by analysing the 477.6 keV gamma energy emitted by the first excited state of ^{7}Li (branching ratio 10.44%) (DDEP 2017) as it achieves the ground state. A typical gamma-spectrometric system consists of a semiconductor crystal detector, liquid nitrogen or mechanical cooling system, preamplifier, detector bias supply, linear amplifier, analogue-to-digital converter (ADC), multi-channel analyser (MCA) of the spectrum, and output data devices (ANSI 1999). An example of a typical gamma detector is given in Fig. 3.1.

High purity germanium (HPGe) detectors currently represent the most widely used gamma-detector systems. The semiconductor crystal is manufactured from ultrapure germanium (impurity level 109 atoms cm^{-3}) in various shapes and in a range of sizes for a wide range of applications (Debertin and Helmer 1988). The specification of a HPGe detector is defined by *(i) the energy resolution, (ii) the detection efficiency* and *(iii) the peak-to-Compton ratio* (see Glossary). When purchasing a HPGe semiconductor detector, these key parameters need to be considered together with the detector material and configuration, detector volume, window material and its thickness. Excepting the ultra-low energy configurations, all detector types commercially available are suitable for the measurement of ^{7}Be activity concentrations.

A. R. Iurian (✉)
Terrestrial Environment Laboratory, IAEA Laboratories Seibersdorf, Seibersdorf, Austria
e-mail: A.Iurian@iaea.org

G. E. Millward
Consolidated Radioisotope Facility (CoRiF), University of Plymouth, Plymouth, UK
e-mail: G.Millward@plymouth.ac.uk

© International Atomic Energy Agency (IAEA) 2019
L. Mabit and W. Blake (eds.), *Assessing Recent Soil Erosion Rates through the Use of Beryllium-7 (Be-7)*, https://doi.org/10.1007/978-3-030-10982-0_3

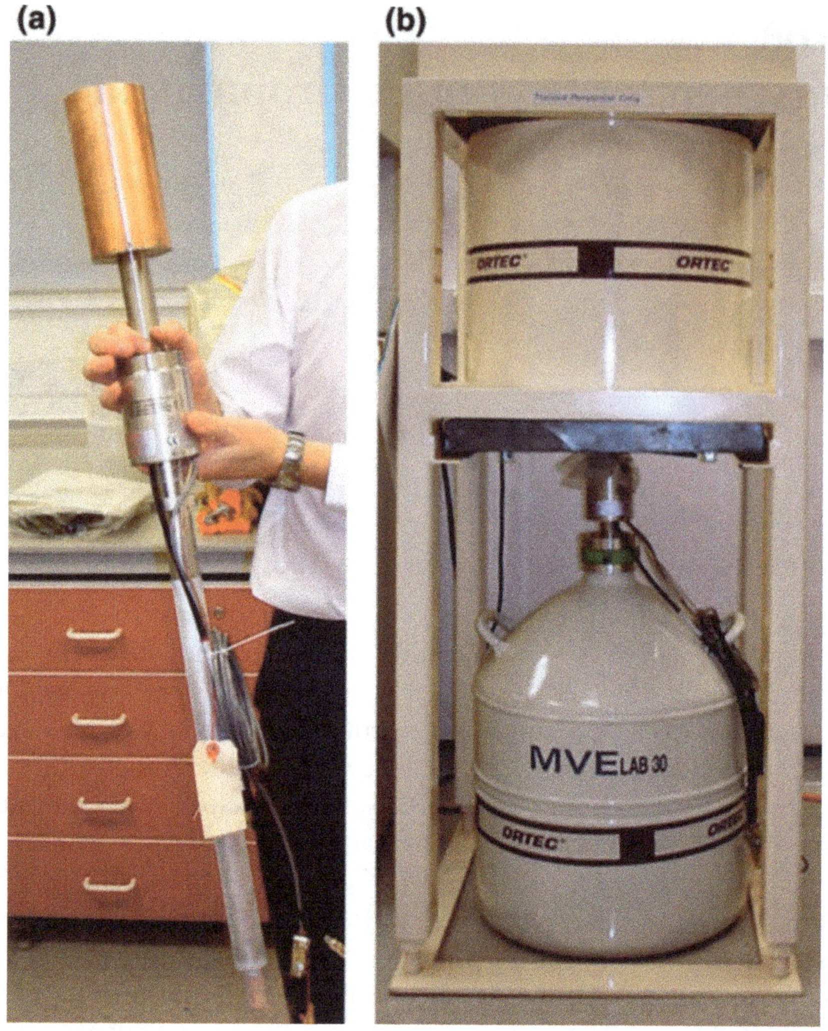

Fig. 3.1 **a** Example of a HPGe gamma detector showing the detector head, the cartridge containing electronic management system and the dip-stick which is immersed in liquid nitrogen to cool down the detector to 77 K; **b** the detector encased in a lead shield with the dip stick immersed in a liquid nitrogen tank. *NB: The liquid nitrogen tank should be filled on a regular basis (every 7–10 days) to maintain a fully operational detector*

3.2 Preparation of Samples

3.2.1 Sample Management

A fundamental consideration for the analyst when managing a set of soil, sediment or rainwater samples is the relatively short half-life of ^{7}Be and its low activity in the majority of environmental materials. It is important, therefore, that the laboratory is well organised and ready to process the samples for ^{7}Be analysis as soon as possible after collection so that a relatively high emission of gamma rays can be taken advantage of. Additionally, maintenance of a high standard of cleanliness in order to prevent contamination of the detector end cap or potential inter-contamination between samples should be prioritized.

3.2.2 Soil Samples

Samples of soil will normally consist of a range of particle sizes. Gamma spectrometric analysis of fallout radionuclides (FRN) in bulk soils should typically be performed on dried, homogenised material that has been sieved to less than 2 mm (Pennock and Appleby 2002) noting that some applications may require sieving to a smaller particle size. Typically, soil samples are oven-dried at 105 °C until a constant weight is reached (commonly for 24 h). A practical alternative is sample freeze-drying, which disaggregates the solids and makes subsequent sample processing more effective. Furthermore, oven-dried samples are disaggregated using a mortar and pestle and size fractionation of the dried solid can then be carried out on the ground sample using a metal or plastic sieve with a defined mesh size. Appendix 3.1 summarises the sample preparation and the associated packing protocol.

3.2.3 Samples of Rainwater

Samples of rainwater from the study site are required to assess the atmospheric deposition flux of ^{7}Be during a period, encompassing the sampling campaign (see Chap. 2). The protocol for extraction of ^{7}Be from rainwater is presented in Appendix 2.1. ^{7}Be can be pre-concentrated from solution by co-precipitation with MnO_2 following the method detailed by Short et al. (2007). The precipitate is recovered on glass fibre filters which are placed in a Petri dish of suitable diameter and allowed to dry before counting on an HPGe gamma detector.

3.2.4 Sample Geometry

Subsequent to physical processing samples, for analysis, can have masses in the range from a few grams to several kilograms, according to the aims of the project and its sample collection requirements. There are a range of measurement geometries, from Marinelli beakers (0.25–1 L) to small vials (~5 cm^{3}) capable of holding a wide range of solid weights (from about 3 g of material to 4 kg) (Fig. 3.2). The choice of the container geometry for ^{7}Be analysis depends on the matrix and on the available amount of material. It is essential that the weights of both empty and filled sample holders are registered on a calibrated laboratory balance to at least 2 decimal places (or at least 4 decimal places for volume calibration standards).

Fig. 3.2 Several examples of containers used to perform gamma spectrometric analysis of solid phases. From left to right: 1 L Marinelli beaker; 300 mL Marinelli beaker; 90 mm diameter plastic Petri dish; 50 mm diameter Petri dish; 4 mL plastic vial (commonly used for conducting analysis using well detectors). To comply with laboratory safety, the radioactive standards are bound in red tape

Prior to purchasing any sample/standard container (especially for Marinelli beaker geometries), the analyst should determine the diameter of the detector end cap so that the container with an appropriate internal diameter will fit as close as possible on the detector end cap. For cylindrical and vial containers, it is recommended to place the samples on a holder (e.g. of light polycarbonate matrix) to assure the reproducibility of the sample-detector arrangement for all samples, including the measurement of the standard. The use of such holders will prevent contamination of the detector end cap, which can be further protected with a thin sheet of cling film or, in the case of well detectors, the sample vial covered in film. Care should be also taken to ensure that the sample/standard container external surface is free of particulate matter.

It is recommended that the geometries of the standard and the soil samples are identical; any differences between them, such as in the sample filling height (e.g. because of insufficient sample amount) or in soil chemical composition, will imply the need for geometry corrections (see Glossary). It is important to note that if the containers are being re-used, the analyst needs to carry out acid cleaning and drying procedures before refilling them. The packing date, nominal weight and an identification code should be indicated on each sample container after packing. As regards filter-mounted samples, these should be measured after placing them in cylindrical Petri dishes of appropriate diameter. Similar contamination precautions are to be taken for the filter geometry as for the other container types discussed above.

3.3 Calibration Approaches

An accurate and precise calibration of a gamma detector is essential for high quality results in the determination of ^{7}Be activity concentrations. The calibration of a gamma spectrometric system involves two major steps:

(1) *Energy calibration* (includes energy calculation as a function of the number of the channel and Full-Width-Half-Maximum (FWHM) as a function of energy);
(2) *Efficiency calibration* (the efficiency as a function of the energy correlated with the source geometry).

Energy calibration is commonly performed using point sources (e.g. ^{152}Eu, ^{137}Cs) or multinuclide standards covering the energy range of interest (Fig. 3.3). The construction of the full energy peak efficiency curve is performed following (1) a *semi-empirical approach* (based on measurement of standard point sources or multinuclide standard sources of soil or liquid matrix), or (2) *Monte Carlo methods* (using a detailed description of the detector and the source). In both cases, the efficiency calibration procedure involves (1) true photopeak efficiency determination for the gamma energies of the radionuclides included in the certified standard source, and (2) the construction of the efficiency curve (within a range of energies given by the radionuclides present in the source) by polynomial fitting. Discrepancies between simulated (Monte Carlo) and experimental efficiency values should be determined and included as an uncertainty component in the uncertainty budget of the ^{7}Be activity value. Note that different detectors and different system-sample configurations will result in different values of the true photopeak efficiencies, which should finally produce the same result of the radionuclide activity value for the actual sample. Coincidence summing corrections should be performed for relevant nuclides present in the calibration source and decaying through a cascade of successive photon emissions (e.g. ^{60}Co, ^{88}Y, ^{139}Ce). These can be determined through general or dedicated codes (Vidmar et al. 2010). Coincidence summing correction factors depend on the nuclide decay scheme, sample geometry and composition, and on detector parameters. The factor is equal to unity in cases where the radionuclide has no cascade of gamma-rays.

3.3.1 Semi-empirical Calibration Method

For routine environmental sample analyses, efficiency calibration of gamma detectors can be performed using (a) multiple standard point sources or standard volume sources of mixed radionuclides (for soils measurements), or (b) filters spiked with standard solution (for ^{7}Be precipitate analysis). Considering the short half-life of ^{7}Be, standard sources containing ^{7}Be are not commercially available. When purchasing a standard for efficiency calibration, it is essential that these should carry a certificate of calibration from a metrological labora-

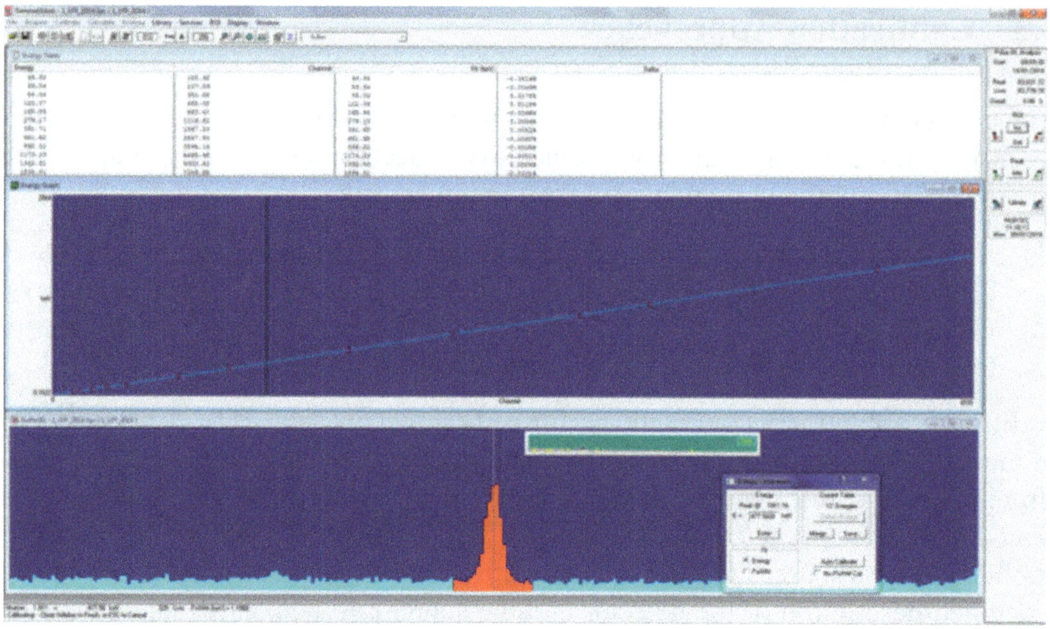

Fig. 3.3 Print screen capture of the gamma detector energy calibration performed using a multi-radionuclide standard and the GammaVision commercial software. Note the interpolated 477.6 keV gamma energy of ^{7}Be between 391.7 keV (^{113}Sn) and 661.7 (^{137}Cs). The upper box is the energy table; the middle box is a plot of the gamma-ray energy against the channel number; the lower box shows the ^{7}Be peak

tory accredited for the preparation of standards. An appropriate calibration standard should have low overall uncertainties of each radionuclide in the source (up to 2–3%), as this uncertainty is further propagated into the uncertainty of ^{7}Be activity. The analyst should be aware that specific international regulations apply for the import of radioactive sources. After purchasing, the calibration standard solution (commonly delivered in sealed glass vial) must be gravimetrically diluted for subsequent use. Furthermore, the analyst shall choose how to use the diluted standard solution for the detector efficiency calibration. Bear in mind that special safety procedures and authorisations apply to the laboratories handling radioactive sources as regards to the risk of radioactive contamination. Appendix 3.2 details a recommended step-by-step approach for the preparation of a multi-nuclide calibration standard of soil matrix.

An example of a multi-nuclide standard solution suitable for ^{7}Be measurements will comprise artificial radionuclides such as ^{241}Am, ^{109}Cd, ^{57}Co, ^{139}Ce, ^{203}Hg, ^{113}Sn, ^{85}Sr, ^{137}Cs, ^{60}Co and ^{88}Y, with energy lines from 59 to 1836 keV. Such multi-nuclide standard can be used to calibrate gamma detectors for efficiency in the vicinity of the gamma emission from ^{7}Be (477.6 keV) which lies between that of ^{113}Sn at 391.7 keV, ^{85}Sr at 514.0 keV (difficult to determine accurately due to the overlapping with the 511 keV annihilation peak) and ^{137}Cs at 661.7 keV (DDEP 2017), as presented in Fig. 3.4. A disadvantage in the short half-lives of some nuclides (e.g. ^{203}Hg, ^{113}Sn)

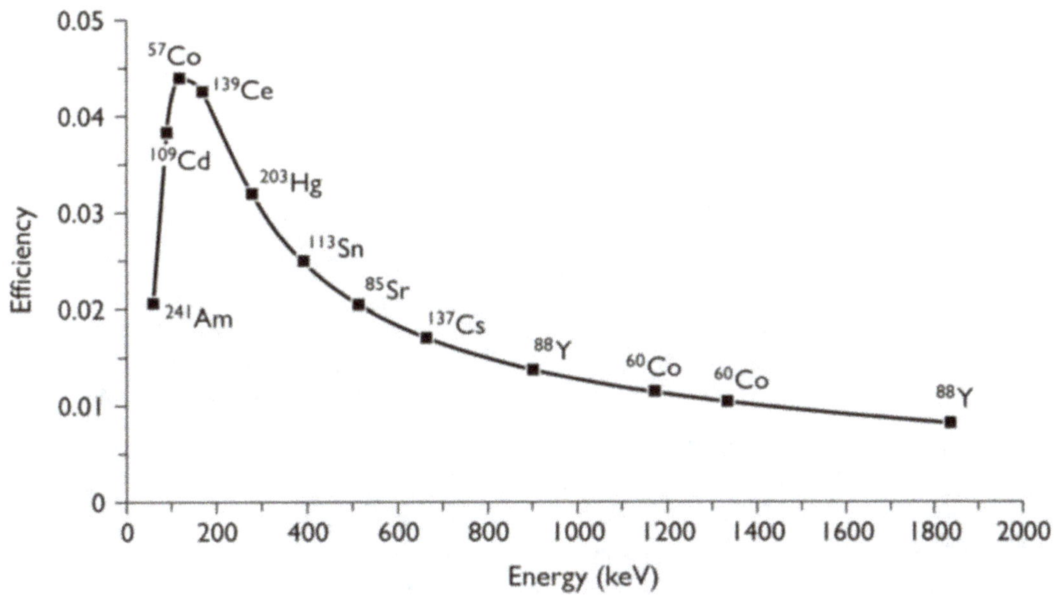

Fig. 3.4 The efficiency calibration curve of a HPGe detector using a multi-nuclide standard solution containing 10 artificial radionuclides (plotted in Microsoft Excel spreadsheet). Note the ^{7}Be gamma emission is interpolated between the gamma energies of ^{113}Sn and ^{85}Sr. After the decay of short-lived radionuclides i.e. ^{85}Sr, ^{113}Sn and ^{203}Hg, the closest lower energy for interpolation will be ^{139}Ce. An interpolation of ^{7}Be energy between ^{139}Ce and ^{137}Cs will led to an increased efficiency value for ^{7}Be

present in the standard must be recognized, thus making the source useful over the whole spectra region only for a short period of time (i.e. a few months).

(i) *Preparation of a secondary soil calibration standard*

A soil sample should be collected, preferably from an area where fallout radionuclides are minimal (e.g. subsoil) and prepared as in Sect. 3.2.2. The detection and quantification of the background gamma radiation in the soil sample is required. Then the prepared soil fraction below 2 mm particle size is spiked with an aliquot of the liquid standard. It is highly important to ensure that the soil-liquid standard mixing is complete, ensuring the spiked material is homogeneous (see step-by-step instruction in Appendix 3.2). The spiked soil is dried and an appropriate sample holder (selected from those in Fig. 3.2), with a similar geometry to the samples, is then selected and filled with the dried spiked soil. The secondary soil standard can then be used to calibrate the detector for efficiency (Fig. 3.5). If there are differences in geometry (different containers and/or filling height) and/or matrix (e.g. different chemical composition) between the soil samples and the calibration source, an efficiency transfer factor should be determined to correct this difference, as ratio between the efficiency of the sample and the efficiency of the standard.

(ii) *Preparation of a secondary liquid calibration standard*

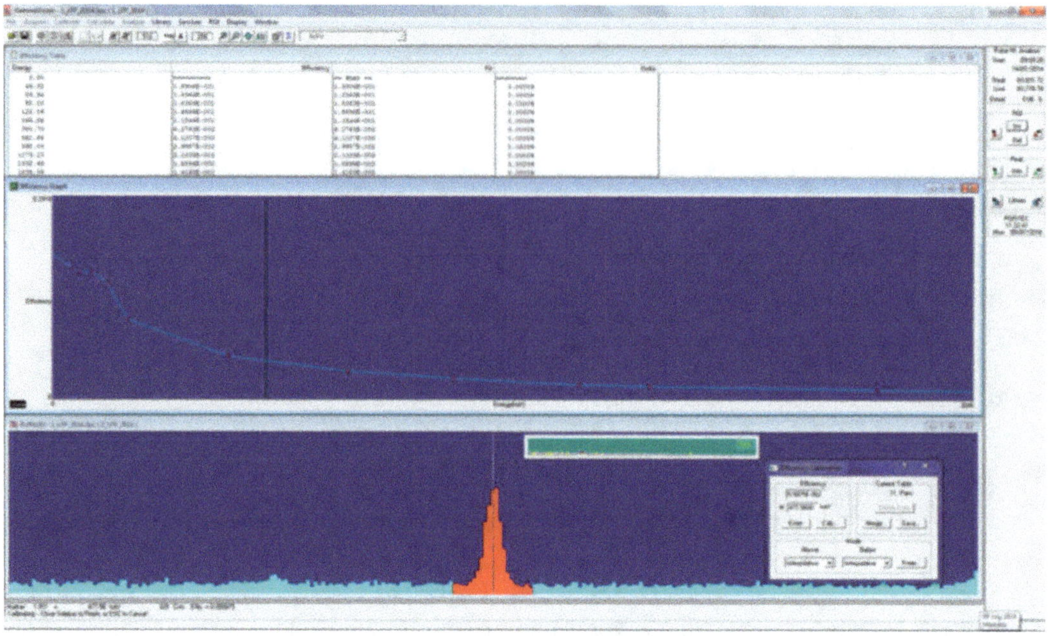

Fig. 3.5 Print screen capture of the gamma detector efficiency calibration performed using a multi-nuclide standard and the GammaVision commercial software. The upper box is the energy table; the middle box is a plot of the efficiencies against the gamma-ray energies; the lower box shows the ^{7}Be peak

In this case, the liquid standard commonly purchased in a glass vial is diluted following the steps in Appendix 3.2, and further poured in different geometries and used as a secondary liquid standard. In this case, the efficiency transfer factor needs to be considered for ^{7}Be activity calculation in soil samples, to account for the difference in matrix between the calibration source and the sample (regardless the sample filling height).

(iii) *Preparation of a secondary filter standard*

A secondary filter standard can be obtained by spiking a filter paper in similar size with the sample (e.g. Whatman filter) with a known content standard solution. This is achieved using a calibrated automatic pipette that is used to add the standard solution dropwise onto the filter. Here the analyst must take care to ensure a homogenous distribution of the standard solution on the filter so that the calibration is coherent with the distribution of precipitate on the sample filter. The spiked filter should then be air-dried prior to sealing in a plastic Petri dish of appropriate diameter.

3.3.2 Monte Carlo Approach

The advanced user may prefer to use Monte-Carlo coding to construct the efficiency curve for a certain sample holder without the need of a certified standard solution.

Monte Carlo approaches for automatic efficiency calibration (e.g. LabSOCS—Laboratory SOurceless Calibration Software) (Bronson and Wang 1996) are commonly integrated into some commercial software of gamma detectors. These codes include several options for detector type and configuration, sample shape and volume (e.g. point, disc, and cylinder), sample/source matrix, source-to-detector distance and for the type of fit used to describe the efficiency-energy dependency (linear, quadratic, polynomial) (Jovanovic et al. 1997; Bronson and Wang 1996). For each particular sample, the analyst needs to create a specific sample geometry based on the holder dimensions (diameter and wall thickness), sample filling height, chemical composition of the sample matrix and density. It is also important for the analyst to know the precise details of the technical characteristics of the gamma detector or to obtain a 'characterisation file' of the detector from its manufacturer in order to run the codes. However, no programming knowledge is necessary to use the available commercial codes. The efficiencies generated for a specific sample can be saved and stored by the analysis software and used to automatically derive the radionuclide activity without performing an experimental calibration. However, the experimentally obtained efficiencies can be as well included in the analysis software, in case these data are available.

The automatic calibration by the Monte Carlo approach eliminates the cost of purchasing, tracking, and disposing of radioactive standards. However, the analytical results still need to be validated using certified reference materials in similar matrices. The main constraints in the application of Monte Carlo approaches are related to uncertainties of the detector configuration (e.g. shape and size of the effective crystal volume, photons and electron interaction probability and angular distributions) and measurement traceability to primary standards.

3.4 Data Handling and ^{7}Be Analysis

Analyses of ^{7}Be can be challenging because of low activity concentrations and, therefore, high counting uncertainties. Sample measurements are completed after performing the energy and efficiency calibrations. Data evaluation is commonly realised using specialised gamma software, provided by the detector manufacturer or made 'in-house'. Prior to spectrum analysis, ^{7}Be nuclear data (gamma energy, half-life and emission probability) should be included in the software library. The stability of the spectrometer should be checked during the measurement and during routine quality assurance procedures. Despite the apparent simplicity of gamma-ray measurements, bear in mind that there are a number of correction factors to the spectrum counting data that must be considered for the determination of ^{7}Be activity concentration e.g. decay correction to the sampling date, efficiency transfer factor, background correction, instrument dead time correction (IAEA 2004), while a correction for soil moisture content is applied to the dried soil sample mass.

A flow chart of ^{7}Be measurement by gamma-ray spectrometry is presented in Fig. 3.6. Sample counting times need to be long enough (typically between 86,000

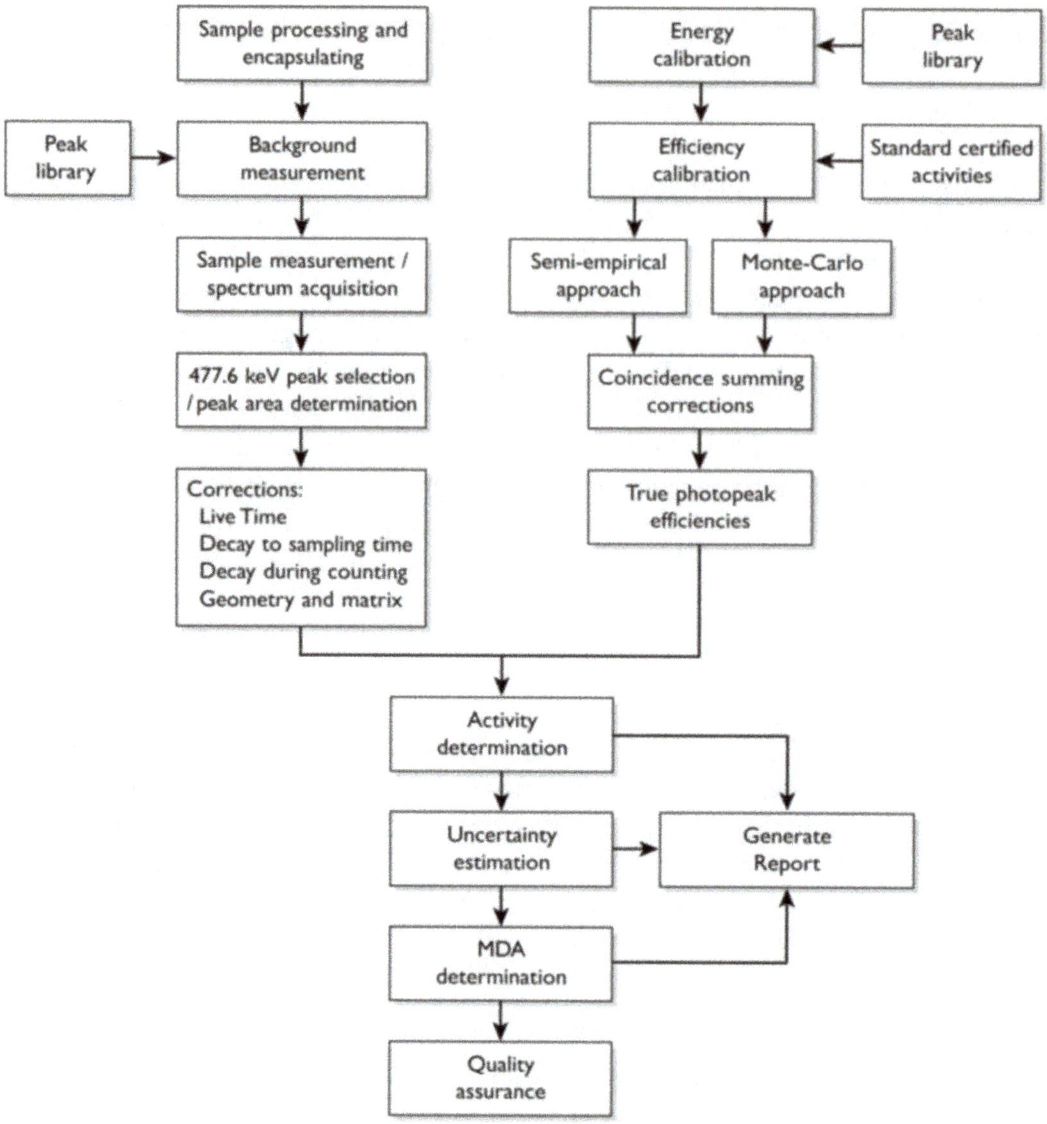

Fig. 3.6 Schematic protocol for ^{7}Be analysis by gamma-ray spectrometry (MDA = Minimum Detectable Activity)

and 259,000 s) to meet laboratory standard statistical uncertainty of the measurement result (commonly between 10 and 30% for ^{7}Be in environmental samples).

Based on the information provided in the library, the instrument-specific gamma evaluation software will identify the ^{7}Be peak and analyse the number of counts within its peak area (Fig. 3.7). Furthermore, following default automated sequences; the software is able to report the activity concentration (Bq kg^{-1}), combined standard uncertainty and minimum detectable activity of the radionuclide found by the peak-search routine (Debertin and Helmer 1988). However, to achieve reliable results using the library approach, any shift effect in the spectrum should be corrected and energy calibration should be precise enough to achieve no more than one or two channels deviation between ^{7}Be reference energy and the one calculated from the calibration curve (IAEA 2002).

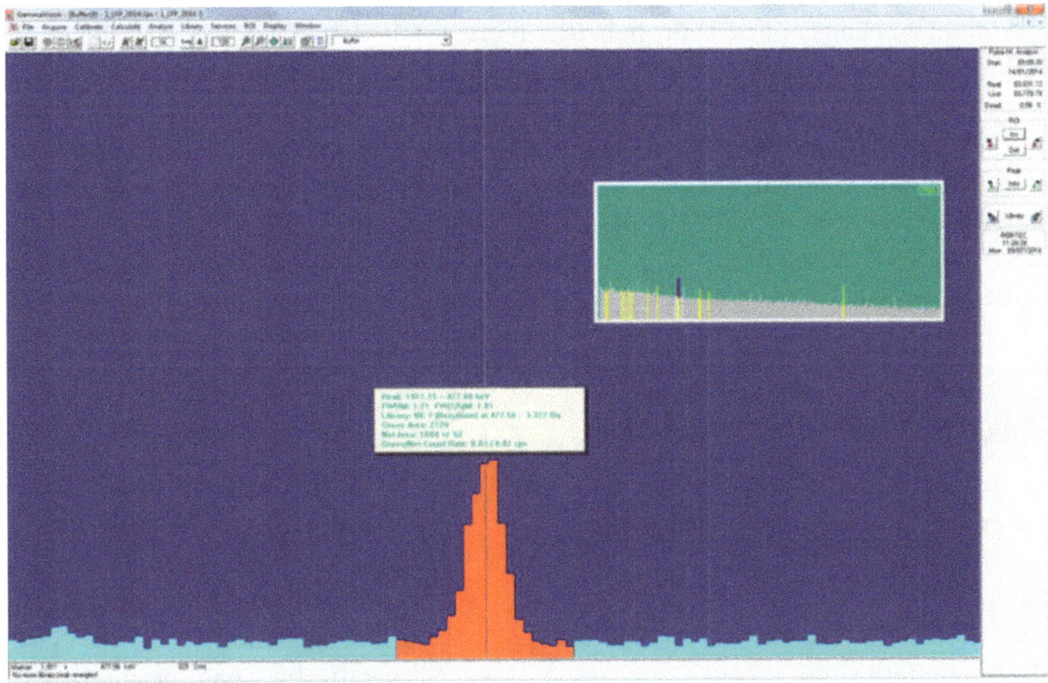

Fig. 3.7 Print screen capture of a gamma spectrum registered with the GammaVision commercial software and showing the ^{7}Be peak at 477.56 keV and its net area and gross/net count rate (peak data box), measurement starting time and live time (upper right part of the screen)

Measurement results are commonly reported with associated combined uncertainty at the 95% confidence level (JCGM 2008). For detailed information on the determination of each uncertainty component, readers can refer to Lépy et al. (2015). The detection limit (LD) provides the detection capabilities of a measurement system.

3.4.1 Quality Assurance

In routine laboratory practice, it is necessary to validate the measurement results for certain working conditions. Both internal and external validation methods are an important part of 'good laboratory practice' and also a requirement of ISO/IEC 17025 for 'in-house' methods (ISO/IEC 17043 2010; ISO/IEC 17025 2017). The internal validation procedure can be employed using distinctive approaches:

- Various types of blanks;
- Replicate analysis to check changes in precision;
- Use of certified reference materials with similar matrix and geometry (e.g. provided by well-recognized international providers such as the IAEA);
- Standards and point sources to check the stability of the response;
- Use of long-term control charts of standards and point source measurements to check the stability of the detector response;
- Proficiency tests participation.

Note that because of its short half-life, there are no reference materials available which contain ^{7}Be. However, it is a common practice to check the accuracy of the laboratory procedure for radionuclides emitting gamma rays in the vicinity of ^{7}Be energy (e.g. ^{137}Cs) in samples of similar matrix and geometry.

In addition, external validation can be achieved through: (i) inter-laboratory comparison exercises and/or (ii) participation in international proficiency tests for radionuclide determination in soil and water samples such as the ones organized regularly by IAEA and its Member States.

Glossary

The energy resolution (FWHM)	The full width of the peak at one-half of the maximum height above any underlying continuous spectral background.[1]
Absolute efficiency	The ratio between the number of pulses recorded and the number of radiation quanta emitted by source.
Relative efficiency	Detector efficiency relative to that of a NaI(Tl) detector.[2]
Peak-to-Compton ratio	The ratio of the number of counts in the biggest channel of the 1332.50 keV ^{60}Co energy peak to the average channel count in the Compton continuum between 1040 and 1096 keV in the same spectrum.
Coincidence summing correction factor	The ratio between the apparent efficiency for the line with energy E of the nuclide having cascading radiations, to the full energy peak efficiency at the same energy obtained from the energy curve measured with single-photon emitting radionuclides. The apparent activity (corrected for background) must be multiplied by the coincidence summing correction factor to

[1] Debertin and Helmer (1988).

[2] Knoll (2000).

	obtain the true activity value for the radionuclides affected by cascade effect.[3]
Geometry and self-attenuation correction factor	The ratio of the full energy peak efficiency for the sample and the full energy peak efficiency for the calibration standard. The correction is commonly performed through an efficiency transfer approach, by multiplying the correction factor with the ^{7}Be full energy peak efficiency.
Efficiency transfer	The "transfer" of the efficiency of a standard to another sample, which can be in the same geometry or not, and acting as corrective factor for both geometry and self-absorption effect.
Minimum Detectable Activity (MDA)	The lowest activity value that can be detected with a measuring system at 95% confidence.[4]
Detection Limit (LD)	The lowest expectation value of the counting rate of a net peak area that can be detected with a measuring system.

Appendix 3.1: Soil Sample Preparation Prior to Gamma Analysis

- Weigh the total wet soil sample in grams to 2 decimal places. Record weight as total wet weight. In case of a soil core, section core into desired slice increments and further follow the same procedure as previously described.
- Freeze dry each soil sample (or oven dry at 105 °C).
- Weight total dry sample (g) to 2 decimal places. Record weight as total dry weight.
- Homogenize each sample with a pestle and mortar to ensure soil material is fully mixed and disaggregated.
- Choose the container size according to the available sample volume to obtain a full (or near full) container after sample packing. Pack a suitable pre-weighed container (Fig. 3.2) with a subsample of the homogenised soil material:

[3]ANSI (1999).

[4]ISO 11929 (2010).

a. Weigh empty container and its lid (g) to 2 decimal places;
b. Pack container with soil and assure that (1) the material is firmly patted down inside the container, so that no void spaces are left within the sample container; (2) the soil surface is flat;
c. Weigh packed container and its lid (g) to 2 decimal places;
d. Record total packed subsample weight (g) to 2 decimal places;
e. Seal container with gas-tight insulating tape;
f. Label container with the pre-arranged sample ID and note date of packing;

- Record all weighed data and sample ID onto supplied spreadsheet.

Appendix 3.2: Preparation of a Secondary Soil Calibration Standard

(A) *Opening and Diluting the Mixed Traceable Standard Solution*

- A mixed traceable standard source is typically purchased from an accredited metrological institute in the form of acid liquid solution inside a flame-sealed glass ampoule. The vial will be opened under fume-hood conditions following the radiation protection laboratory procedures.
- Caution should be taken to avoid any spillage of the solution when the 'swan neck' is broken (along the scratch mark).
- A clean, acid-washed 100 mL Grade A glass volumetric flask is weighed (4 decimal places) and fitted with a small funnel. Most of the traceable reference source is transferred to the volumetric flask using a disposable Pasteur pipette.
- The Pasteur pipette is then filled with 1–2 mL of 4 M HCl which is used to wash the remaining traceable solution into the volumetric flask via the funnel. This should be repeated several times to ensure that all the traceable solution is transferred to the volumetric flask. When all the liquid has been transferred from the vial the liquid in the flask should be filled up to the mark with 4 M HCl and weighed again for 3 times. The mean arithmetic weight of the flask with the standard solution is recorded to 4 decimal places.
- The liquid in the volumetric flask is then transferred to a clean plastic bottle and labelled as 'Stock Solution' of the mixed radionuclide reference source and clearly marked with trefoil.

(B) *Dosing a Soil Sample with the Mixed Standard Solution*

- A soil sample, typically about 100 g, with the grain size <2 mm is prepared as mentioned in Sect. 3.3.1(i) and placed in a suitable beaker. The soil is moistened with a small volume of distilled water until the water level is above the soil surface, to enable the mixing of the slurry with the liquid standard.
- An appropriate volume of the diluted mixed traceable reference solution is added to the slurry using a calibrated pipette. The amount added should be sufficient to

achieve <1% uncertainty of the counting statistics of the peaks of interest within a reasonable counting time when measuring the calibration standard.

- The slurry is thoroughly mixed using a glass rod.
- The slurry is dried in an oven at 105 °C or, preferably, in a freeze-drier until constant weight is achieved (commonly over 24–48 h).
- The dried solid is then carefully transferred to a rotating ball mill to ensure that the final dried solid is completely mixed. The final product is a fine powder that can be allocated to appropriate sample holders, such as 50 mm diameter, dried, pre-weighed Petri dishes.

For safety, any particle transfer process should be completed in a fume hood and the operator should be protected with a mask.

- The filled Petri dish, and any other containers, should be re-weighed and sample weights recorded so the activity concentrations of each of the radionuclides in the standard can be evaluated. Afterwards, the standard should be bound with adhesive tape in distinct colour (e.g. red) for safety reasons. Its sample number, or reference data, should be written on the lid in indelible ink.
- The Petri dish should then be wiped with a moist cloth to remove any extraneous particles. It should be kept in a sealable plastic bag when out of use.
- These will be further used as secondary standards, in different geometries, for the efficiency calibration of the gamma detectors. A record should be kept on the radionuclide inventory: (i) present in the secondary standard, and (ii) left in the 'Stock Solution'.

References

ANSI National Standards Institute. (1999). ANSI N42.14—1999, Calibration and use of germanium spectrometers for measurement of gamma-ray emission rates of radionuclides. New York: IEEE, Inc.

Bronson, F.L., & Wang, L. (1996). Validation of the MCNP Monte Carlo Code for germanium detector efficiency calibration. In *Proceedings of the Waste Management 1996 Congress, Tuscon AZ, USA*.

Debertin, K., & Helmer, R. (1988). *Gamma- and X-ray spectrometry with semiconductor detectors* (367 p). North-Holland, Amsterdam.

Decay Data Evaluation Project (DDEP). (2017). Recommended Data. http://www.nucleide.org/DDEP_WG/DDEPdata.htm.

IAEA. (2002). *Specialized software utilities for gamma ray spectrometry Final report of a co-ordinated research project 1996–2000* (100 p). IAEA-TECDOC-1275. Vienna, Austria: International Atomic Energy Agency Publication.

IAEA. (2004). *Quantifying uncertainty in nuclear analytical measurements* (250 p). IAEA-TECDOC-1401. Vienna, Austria: International Atomic Energy Agency Publication.

ISO/IEC 17025. (2017). *General requirements for the competence of testing and calibration laboratories*. Geneva: ISO.

ISO 11929. (2010). *Determination of the characteristic limits (decision threshold, detection limit and limits of the confidence interval) for measurements of ionizing radiation - Fundamentals and application*. Geneva: ISO.

ISO/IEC 17043. (2010). *Conformity assessment - General requirements for proficiency testing*.

JCGM. (2008). JCGM 100:2008. GUM 1995 with minor corrections. Evaluation of measurement data—Guide to the expression of uncertainty in measurement. http://www.bipm.org/en/publications/guides/.

Jovanovic, S., Dlabac, S., Mihaljevic, N., & Vukotic, P. (1997). ANGLE—a PC—code for semiconductor detector efficiency calculations. *Journal of Radioanalytical and Nuclear Chemistry, 218,* 13–20.

Knoll, G. F. (2000). *Radiation detection and measurement* (3rd ed., 802 p). Wiley.

Lépy, M. C., Pearce, A., & Sima, O. (2015). Uncertainties in gamma-ray spectrometry. *Metrologia, 52*(3), S123–S145.

Pennock, D. J., & Appleby, P. G. (2002). Chapter 4. Sample processing, In F. Zapata (Ed.),*Handbook for the assessment of soil erosion and sedimentation using environmental radionuclides* (pp. 59–65). The Netherlands: Kluwer Ac. Publ.

Short, D. B., Appleby, P. G., Hilton, J. (2007) Measurement of atmospheric fluxes of radionuclides at a UK site using both direct (rain) and indirect (soils) methods. *International Journal of Environment and Pollution*, *29*(4), 392–404.

Vidmar, T., Celik, N., Cornejo Diaz, N., Dlabac, A., Ewa, I. O. B., Carrazana Gonzalez, J. A., et al. (2010). Testing efficiency transfer codes for equivalence. *Applied Radiation and Isotopes, 68,* 355–359.

The opinions expressed in this chapter are those of the author(s) and do not necessarily reflect the views of the International Atomic Energy Agency, its Board of Directors, or the countries they represent.

Open Access This chapter is licensed under the terms of the Creative Commons Attribution 3.0 IGO License (https://creativecommons.org/licenses/by/3.0/igo/), which permits use, sharing, adaptation, distribution and reproduction in any medium or format, as long as you give appropriate credit to the International Atomic Energy Agency, provide a link to the Creative Commons licence and indicate if changes were made.

The use of the International Atomic Energy Agency's name, and the use of the International Atomic Energy Agency's logo, shall be subject to a separate written licence agreement between the International Atomic Energy Agency and the user and is not authorized as part of this CC-IGO licence. Note that the link provided above includes additional terms and conditions of the licence.

The images or other third party material in this chapter are included in the chapter's Creative Commons licence, unless indicated otherwise in a credit line to the material. If material is not included in the chapter's Creative Commons licence and your intended use is not permitted by statutory regulation or exceeds the permitted use, you will need to obtain permission directly from the copyright holder.

Chapter 4
Conversion of Be-7 Activity Concentrations into Soil and Sediment Redistribution Amounts

W. H. Blake, A. Taylor, A. R. Iurian, G. E. Millward and L. Mabit

4.1 Implementing the Event-Scale Profile Distribution Model

The Profile Distribution Model (PDM) is described in detail by Walling and He (1999) and Blake et al. (1999). Here, alongside an overview of the basic components, we provide step-by-step guidance to its implementation and simple encoding within Microsoft Excel.

This simple conversion model relies on exponential decline in ^{7}Be activity concentration with depth wherein the shape of the profile is described by h_0 (See Chap. 1, Sect. 1.2). A first necessary step is to establish a specific h_0 value for the site under investigation. This can be achieved by plotting the depth profile from your reference area as an x–y plot and fitting an exponential function (see Table 4.1 and Fig. 4.1).

W. H. Blake (✉)
School of Geography, Earth and Environmental Sciences, University of Plymouth, Plymouth, UK
e-mail: william.blake@plymouth.ac.uk

A. Taylor · G. E. Millward
Consolidated Radioisotope Facility (CoRiF), University of Plymouth, Plymouth, UK
e-mail: alex.taylor@plymouth.ac.uk

G. E. Millward
e-mail: G.Millward@plymouth.ac.uk

A. R. Iurian
Terrestrial Environment Laboratory, IAEA Laboratories Seibersdorf, Seibersdorf, Austria
e-mail: A.Iurian@iaea.org

L. Mabit
Soil and Water Management and Crop Nutrition Subprogramme, Joint FAO/IAEA Division of Nuclear Techniques in Food and Agriculture, International Atomic Energy Agency, Vienna, Austria
e-mail: L.Mabit@iaea.org

© International Atomic Energy Agency (IAEA) 2019

L. Mabit and W. Blake (eds.), *Assessing Recent Soil Erosion Rates through the Use of Beryllium-7 (Be-7)*, https://doi.org/10.1007/978-3-030-10982-0_4

Table 4.1 Example of the depth profile data used in Fig. 4.1 where the ^{7}Be concentrations have been determined with a 2-sigma error

Nominal true depth (mm)	Total dry mass (<2 mm) (g)	Mass depth (kg m^{-2})	Cumulative mass depth (kg m^{-2})	^{7}Be activity concentration (Bq kg^{-1})	± (2 sigma)[a]
2	83.84	0.93	0.93	70.02	4.5
4	76.17	0.85	1.78	61.85	5.0
6	90.14	1.00	2.78	46.91	5.2
8	84.71	0.94	3.72	52.51	5.2
10	165.91	1.84	5.56	48.06	4.4
12	156.87	1.74	7.31	48.62	6.2
14	175.54	1.95	9.26	24.89	5.0
16	204.27	2.27	11.53	19.52	5.7
18	162.85	1.81	13.34	11.03	5.4
20	214.66	2.39	15.72	10.13	5.2

[a]Analytical uncertainty

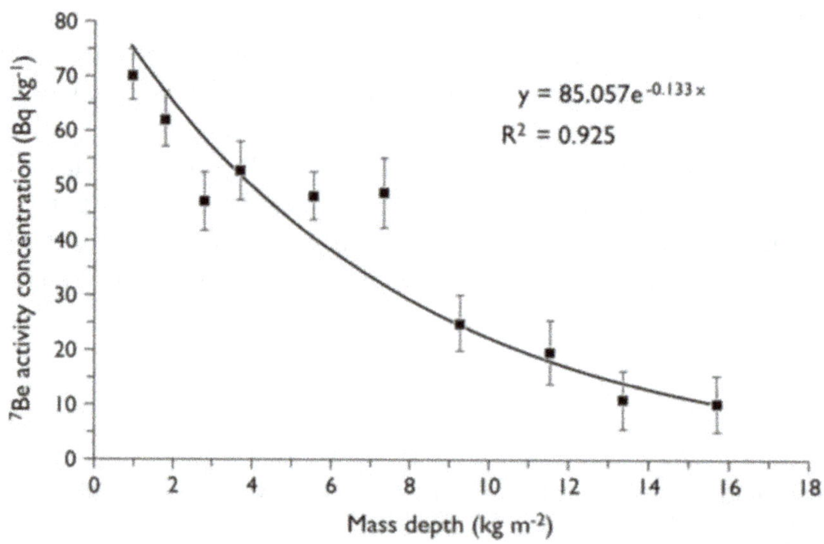

Fig. 4.1 ^{7}Be activity concentration against mass depth with a fitted exponential line to derive the value of h_0

It is important that measures of depth are represented as mass depth i.e. mass of soil for the core slice divided by the area of the core with units of kg m^{-2}. This accounts for changes in bulk density and is a more precise measure of depth than true depth. h_0 is derived by taking the reciprocal of the exponent of the line. In the case of the depth profile shown in Fig. 4.1, h_0 is derived from:

$$\frac{1}{0.133} = 7.5\left(\text{kg m}^{-2}\right)$$

Table 4.2 An example grid cell layout with example inventory data (±is analytical uncertainty) where coding presented in Table 4.3 is required to determine h, C_d and h' (see Chap. 1, Appendix 1.1), without particle size correction. Calculated values following Table 4.3 are shown in italics

	A	B	C	D	E	F	G
1	Sample ID	^{7}Be inventory (Bq m^{-2})	±	Uncorrected erosion amount (h) (kg m^{-2})	Eroded soil ^{7}Be activity conc. (Bq kg^{-1})	Deposited soil ^{7}Be activity conc, (C_d) (Bq kg^{-1})	Uncorrected deposition amount (h') (kg m^{-2})
2	1	184	38	*7.85*	*43.32*		
3	2	394	56	*2.14*	*60.79*		
4	3	300	43	*4.18*	*53.55*		
5	4	216	34	*6.65*	*46.34*		
6	5	304	52	*4.08*	*53.88*		
7	6	279	45	*4.73*	*51.83*		
8	7	361	101	*2.79*	*58.33*		
9	8	397	103	*2.08*	*61.01*		
10	9	330	69	*3.47*	*55.94*		
11	10	435	98	*1.40*	*63.75*		
12	11	817	102			*51.82*	*5.65*
13	12	1070	96			*51.82*	*10.54*

To run the PDM in Excel, the key components of the model can be coded into a cell grid as illustrated in Table 4.2. In this example, the reference inventory value is set at 524 ± 46 Bq m^{-2} (n = 15 where uncertainty represents 2 standard deviations). The sample points 1–10 are erosional (i.e. an inventory deficit as compared to the activity recorded in the reference area) and sample points 11 and 12 are depositional (i.e. an inventory excess as compared to the activity recorded in the reference area).

Coding required for columns D, E, F and G to calculate h, C_d and h' respectively (see Chap. 1, Appendix 1.1 for relevant model equations) is provided in Table 4.3.

The PDM may also be applied using a specific Microsoft Excel Add-in—developed by Walling and He (1999) and refined by Walling et al. (2002)—which is available at the IAEA website (see http://www-naweb.iaea.org/nafa/swmn/models-tool-kits.html).

4.2 Accounting for Size Selectivity of Erosion and Deposition Processes in the Profile Distribution Model

As pointed out in the Sect. 2.4 of the Chap. 2, some advanced users will need to account for particle size selectivity associated with soil erosion processes to convert inventory loss into soil erosion amounts. Different approaches are described in the literature (see Taylor et al. 2014 or Yang et al. 2013). Here, we describe the method

Table 4.3 Coding for columns D, E, F and G in Table 4.2 to derive the key elements of the PDM as described in Sect. 1.2

Cell number (assuming layout as per Table 4.2)	Code	Explanation
D2	=\$I\$2 ∗ LN(\$I\$3/B2) (see Chap. 1 Eq. 1.3)	This Excel code is to derive 'h' i.e. the mass depth of soil lost (NB as positive value) i.e. *Aref/A* to account for the observed inventory deficit in cell D2 **where cell I2 contains the value for h_0 (in this example 7.5 kg m^{-2}) and cell I3 contains the reference inventory (in this case A_{ref} = 524 Bq m^{-2})**. Note that this code can then be copied down to all eroding site points
E2	=(\$I\$3 − B2)/D2 (see Chap. 1 Eq. 1.5)	This Excel code permits to estimate the ^{7}Be concentration of the soil eroded from this sample point or zone
F11	=(D2 ∗ E2)+…+(D11 ∗ E11))/ SUM(D2:D11) (see Chap. 1 Eq. 1.6)	This Excel code allows to calculate the mean activity concentration of upslope eroded soil used in the equation for *h′* below. Note that all upslope contributing points need to be included in the code across (shown in shortened form). This needs to be copied for all depositional points
G12	=(B12 − \$I\$3)/F12 (see Chap. 1 Eq. 1.4)	This is to calculate h′ i.e. the mass depth of soil gained to account for the inventory excess in cell D11. This needs to be copied for all depositional points

proposed by Taylor et al. (2014) and compare to other methods so users can make an informed decision about which approach to take.

The PDM assumes that the soil profile is top-sliced by erosion and that loss of inventory is directly related to the mass depth of soil lost (see Fig. 1.3). However, enrichment of ^{7}Be in fine soil fractions (Fig. 4.2) means that these can be preferentially removed by erosion processes resulting in an augmented inventory deficit relative to erosion amount (Blake et al. 2009; Taylor et al. 2014). It should be noted that in some cases where soil has an overall fine texture and rainfall is of high magnitude, particle size correction may be less important (e.g. Porto and Walling 2014)

The approach suggested by Taylor et al. (2014) accounting for particle size selectivity follows the principles of the method proposed by He and Walling (1996) wherein particle size correction factors P and P′ for eroding and depositional sites respectively are calculated based on measurements of 'geometric' Specific Surface Area (SSA unit in m^2 g^{-1}). P is derived from:

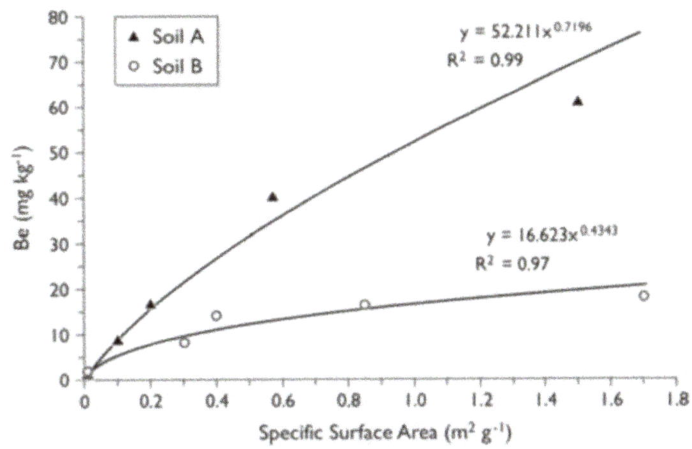

Fig. 4.2 Enrichment of stable Be, as a surrogate for ^{7}Be, in two fine grained soil materials with contrasting texture where soil A has a greater sand content (Taylor et al. 2014)

$$P = \left(\frac{S_m}{S_s}\right)^v \tag{4.1}$$

where S_m is the mean SSA of the mobilised fraction and S_s is the mean SSA of the bulk soil material and v is a constant describing the relationship (i.e. the power function exponent) between ^{7}Be concentration and SSA (Fig. 4.2).

P' is derived from:

$$P' = \left(\frac{S_d}{S_m}\right)^v \tag{4.2}$$

where S_d is the mean SSA of the deposited material.

The SSA of soil and sediment samples is most commonly performed using laser granulometry. It is also possible to derive SSA from particle size data derived by gravitational sedimentation methods, most commonly the pipette method (Day 1965).

It is important at this stage to consider that amounts of erosion estimated using a conversion model consider *inventory deficit* i.e. the amount of ^{7}Be inventory lost from a specific area due to soil erosion. This inventory deficit is converted to mass depth soil loss using the exponential shape of the ^{7}Be distribution in the soil profile (Eq. 1.1). In this context, Taylor et al. (2014) propose that when correcting erosion rates (mass depth lost, kg m^{-2}) for particle size enrichment it is, therefore, necessary to apply the correction factor to the *inventory deficit* only (i.e. A minus A_{ref}) which has the effect of suppressing the loss in inventory according to the particle size of the fraction mobilised. If fine sediment is preferentially mobilised by erosion processes, failure to account for this will lead to overestimation of erosion.

In the case of the above, following the principles of He et al. (2002), a particle-size corrected erosion rate, h_c, can be calculated using the following equation (compare to Eq. 1.2):

$$h_c = h_0 \ln \frac{A_{ref}}{A_{ref} - (A_{ref} - A)/P} \tag{4.3}$$

For estimates of deposition, the ^{7}Be concentration of deposited material needs to be adjusted according to the particle size of the deposited fraction relative to the eroded fraction. The equation to provide corrected deposition rates, h'_c, is (compare to Eq. 1.4):

$$h'_c = \left(A - A_{ref}\right)/C_{d,c} \tag{4.4}$$

where $C_{d,c}$ is the weighted mean concentration of sediment mobilised from the upslope contributing area (Blake et al. 1999; Walling et al. 1999) which has been corrected for particle size enrichment during erosion, and then corrected for depletion during deposition (He et al. 2002). The particle size corrected concentration of deposited material from an individual eroding point (cd') can be estimated using the following function:

$$cd' = PP'A_{ref}\left(1 - e^{-h/h_0}\right)/h \tag{4.5}$$

These data processing requirements need to be considered in the context of the sampling challenges detailed in Chap. 2.

The above equations can be used to modify the Excel-based PDM model coding provided in Table 4.3.

There are several challenges with particle size correction that all users must be aware of. Determining the relationship between the activity concentration and SSA to gain the value of the exponent, v, in Eqs. (4.1) and (4.2), requires particle size separation experiments (Taylor et al. 2014) and use of a laser granulometer or quantitative sieving (and application of SSA theory which makes assumptions about particle sphericity). Sampling for S_s and S_d is relatively straightforward (sampled in field from stable and deposition sites) but sampling for S_m is less straightforward. Options for this include rainfall simulation experiments on the hillslope after the erosion event to collect representative eroded material, installation of Gerlach troughs in study field prior to anticipated erosion events or estimating SSA of eroded material from remaining soil-and comparison with uneroded material (see Yang et al. 2013)

All particle size corrections require a necessary simplification of reality since soil characteristics are highly variable within and between sites. Example data illustrating the influence of particle size correction approaches on soil erosion estimate results are given in Sect. 4.4.

4.3 Extended Time Series Conversion Model: The Theory

If an advanced user wishes to apply the ^{7}Be approach over a longer time period, for example over a wet season that contains notable erosional events in series, an alternative approach is required (Walling et al. 2009). Modifications to the PDM approach are required to account for inventory flux (Bq m^{-2}), decay and the erosivity of events.

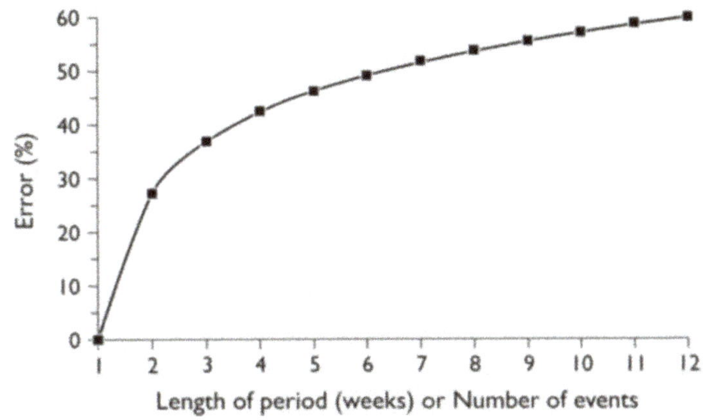

Fig. 4.3 Relationship between the percentage under-estimation of the erosion rate versus the length of the study period (from Walling et al. 2009)

Failure to account for these factors has been shown to lead to underestimation of redistribution rates (Walling et al. 2009). In this approach, the cumulative change in inventory at the sample site is considered rather than a direct comparison to the stable reference value for each day. A reference inventory for the field site is required, however, to provide an indication of erosion or deposition at each individual sample location.

The basic principle of the extended time-series model is that the amount of erosion estimated by ^{7}Be inventory deficit and deposition estimated by ^{7}Be inventory excess needs to be apportioned across the time period of study while allowing for the effects of radioactive decay. The problem of the PDM over extended time periods is considered by Walling et al. (2009) for a hypothetical scenario where: (1) the reference inventory begins at zero due to extended dry period, (2) a single rainfall event of same intensity takes place each week with same erosivity, (3) h_0 remains the same. The results (Fig. 4.3) demonstrate that the ^{7}Be approach increasingly underestimates the erosion rate with substantial error as time progresses into the study period (e.g. 30% after 2–3 weeks; 50% after 3 months). In addition to this issue of decay of the earlier erosion event signals, there are other potential complications linked to temporal distribution of ^{7}Be during the study period, temporal distribution of erosion (rainfall erosivity), potential variability of other key parameters through time, especially on freshly ploughed surface, e.g. h_0.

In this context, every user needs to consider the sensitivity of their study data to these issues and either take action or acknowledge these limitations and the uncertainty introduced into their results.

In this section we present a step-by-step approach to implementing the extended time series model in Microsoft Excel based on Walling et al. (2009). There are three main components: (1) calculating relative erosivity of rainfall, (2) estimating soil loss and (3) estimating deposition.

The aim of the extended time series model is to allow ^{7}Be measurements to be extended over a longer timeframe that is more representative for some environments for example where erosion takes place over a prolonged wet season.

In addition to the evidence collected for the PDM, the extended time series model requires a reconstructed record of daily fallout deposition (Bq m^{-2} d^{-1}) for the whole study period and a time series of daily (where appropriate) relative erosivity (see below). Together, these account for the effects of time variant erosion and the time variant fallout as well as for the radioactive decay.

The first step in applying the model is the same as for the PDM (Sect. 4.1) namely to determine whether the site is erosional or depositional i.e. compare inventory at erosional point (A) or depositional point (A') to reference inventory A_{ref}.

The second step is to build a mass balance model framework used to define the ^{7}Be inventory at the end of each day $A(t)$ (Bq m^{-2}). This is determined from: (i) the ^{7}Be inventory of the previous day $[A(t-1)]$, (ii) effects of radioactive decay [λ], (iii) any fallout input (that day [$F(t)$], (iv) any loss in ^{7}Be inventory due to erosion [A_{loss}]. Of course, the mass balance must ultimately be solved for the latter element. These components are linked together as follows:

$$A(t) = A(t-1)\exp(-\lambda) + F(t) - A_{loss}(t) \tag{4.6}$$

A_{loss} (t) in Eq. (4.6) will reflect (i) amount of soil eroded [$R(t)$] as a mass depth i.e. kg m^{-2}, (ii) existing inventory at the point [$A(t)$], and (iii) depth distribution of inventory i.e. h_0 (for that day) i.e.:

$$A_{loss}(t) = \left[A(t-1)\exp(-\lambda) + F(t)\right] * \left[1 - \exp\left(-\frac{R(t)}{h_0(t)}\right)\right] \tag{4.7}$$

In the prior equations, the amount of soil eroded $R(t)$ is assumed to be proportional to the rainfall erosivity for that daily time-step and can be expressed as following:

$$R(t) = E_r(t) * C \tag{4.8}$$

In Eq. (4.8), C is a constant that links relative erosivity (i.e. E_r) to erosion amount. The value of C differs between sample points but at any one sample point it remains constant through the study period.

To derive C and hence $R(t)$, the user must establish a continuous mass balance for the study period for each sampling point experiencing erosion. Guidance on setting this up in Microsoft Excel is given in the next section. The value of $A(t)$ at the end of the study can then be related to the measured inventory (A) at the time of sampling. Note the value at the start is equal to the measured reference inventory at the beginning of the study (often zero due to cultivation of extended dry period). Given all the data that have been derived up to this point, the mass balance can be solved for C using the optimising Solver routine in Microsoft Excel. This so called 'what-if analysis' tool basically finds the optimum value of C to explain the measured inventory within the constraints of the mass balance detailed in the worksheet.

Application of C to the $E_r(t)$ values in the time series then gives the estimated erosion amount for the sample point. An example of spreadsheet coding is given in Sect. 4.4.

As noted above, in the extended time-series model, a key part of the process is using rainfall erosivity to apportion erosion potential across the study period. For the study period in question, a measure of relative erosivity is required for each event occurring during the study period.

Arguably, we need knowledge of site specific thresholds for erosion initiation based on rainfall patterns and dynamics and site-specific conditions need consideration e.g. runoff initiation processes. Walling et al. (2009) used high resolution rainfall data (minimum 30 min interval data) to evaluate temporal distribution of erosivity (E_r). This approach is linked to RUSLE daily values of erosivity based on the product of the total kinetic energy of rainfall (E) and maximum 30 min rainfall intensity (I_{30}) i.e. EI_{30}. This means a continuous rainfall record at study site is required (30 min totals). It should be noted that a measure of relative erosivity in this context is more important than an absolute value. Local empirical data, if available, may be more appropriate at some sites. Taylor (2012) followed procedures outlined by Yin et al. (2007) and Morgan (2001) to derive EI_{30} following Schuller et al. (2010).

4.4 Implementing the Extended Time Series Model

The extended time series model can be implemented in a variety of ways. We propose here one version for coding in Microsoft Excel as an example to develop from. Such an approach permits the user to see the model working step-by-step and to troubleshoot any coding errors or data anomalies that emerge.

The model structure comprises a series of columns that represent the time series of daily fallout, erosion etc. (Table 4.4) and a set of 'model interface' cells where primary data is inputted and results are calculated.

To permit cell O7 to show the most likely erosion rate, the Solver Routine must be run. This requires the actual measured inventory at the sample point to be entered into the Solver dialogue box (Fig. 4.4) and the routine run so the value of O5 is adjusted to equal the measured inventory. The routine does this by adjusting the unknown constant C until the value of O7 is equal to the inventory measured.

The extended time series model is an appropriate approach to address the challenge of using ^{7}Be across multiple rainfall events. In addition to the requirements of the PDM, it requires rainfall data and site-specific determination of relative erosivity. Implementation of this advanced model requires careful coding and usage of data. Users need to be aware of specific sample point attributes and relationships to other points and to ensure that each working step is traceable in case of a need to revisit data.

Table 4.4 Microsoft Excel coding for time series and mass balance columns. It is assumed user has a basic knowledge of Microsoft Excel and that the column headings are entered in the first row

Cell number	Code	Column title and explanation
A2 (and relevant cells below)	n/a	Date of time step (in DD/MM/YYYY format). This should be copied down for the duration of the study period from the start of rainfall measurement to the final erosion event to be studied
B2 (and relevant cells below)	n/a	Daily inventory fallout flux (F) value (Bq m^{-2}) which has been derived from a rainfall measurement and sampling programme described in Chap. 2
C2 (and relevant cells below)	=0.0130	Decay constant for ^{7}Be for 1 day time step
D2	n/a	The study area initial plot reference inventory at the start of the rainfall measurement and monitoring, typically zero (Bq m^{-2})
D3	=B3 + (D2 * EXP(−C3 * 1))	Daily reference inventory for time step based on decay corrected inventory from prior day plus the fallout input of the current day (Bq m^{-2}). Note this column is copied down to the end of the study period
E2 (and relevant cells below)	n/a	Relative Erosivity (E_r) value for the timestep. Note that all E_r values in the column will sum to 1. Note this column is copied down to the end of the study period
F2	=$O$2	Constant C. Note this code links it to a cell in the model 'interface' cells. Note this column is copied down to the end of the study period
G2	=E2 * F2	Relative Erosion Rate (R). Note this column is copied down to the end of the study period
H2 (and relevant cells below)	=G2/$O$3	Particle size corrected relative erosion rate where cell O3 contains the simple correction factor P in the model 'interface' cells area (based on Yang et al. 2013). See Sect. 4.2. Note this column is copied down to the end of the study period
I2	=(D2) * (1 − EXP(G2/$O$4))	Inventory loss (A_{loss}) (Bq m^{-2}) where cell O4 contains the value of h_0. This is for first time step. For following time steps, cell I3 then needs updating as below to account for decay or previous time-step
I3	=(J2 * EXP(−C3) + B3) * (1 − EXP(G3/$O$4))	Note this column is copied down to the end of the study period
J2	=D2 − I2	Modelled inventory (A) (Bq m^{-2}) for first time step. For following time steps, see cell J3 as below to account for decay or previous time-step
J3	=J2 * EXP(−C3 * 1) + B3 − I3	Note this column is copied down to the end of the study period
K2	=IF(G2 <>0, I2/G2, 0) * $O$3	Activity of eroded soil (C_e) (Bq kg^{-1}). This is used in deposition calculations downslope

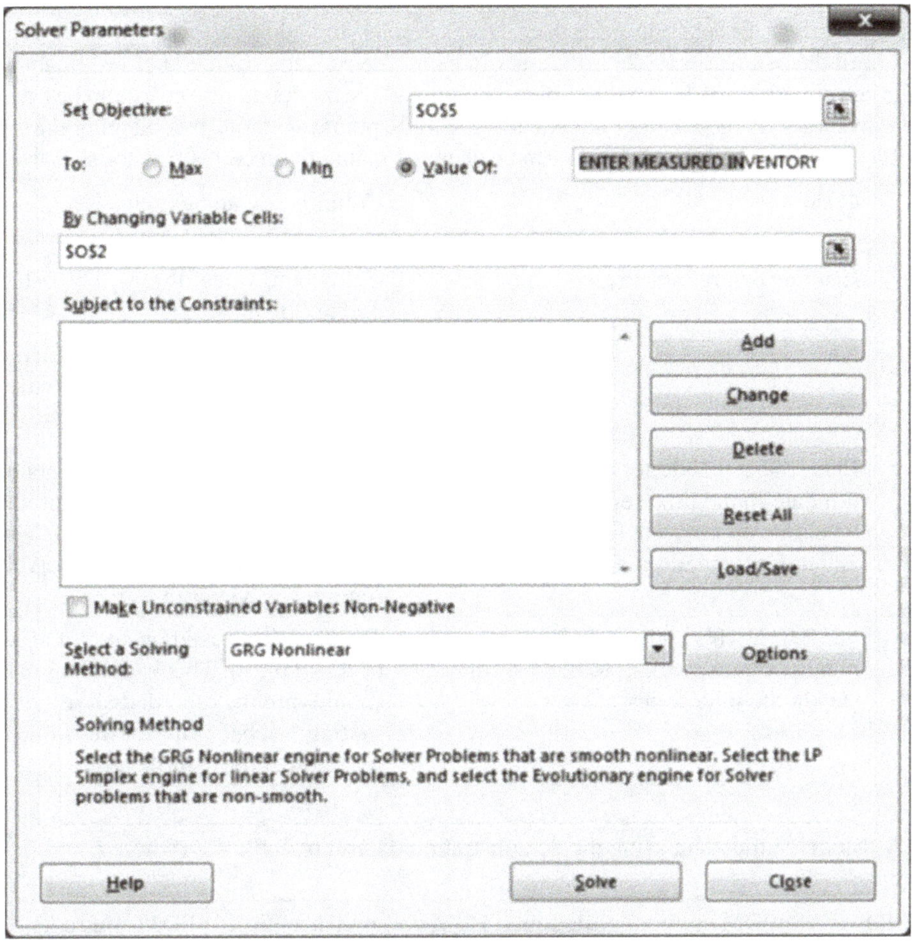

Fig. 4.4 Dialogue box for Miscrosoft Excel Solver command set up to work with the user interface cells in Table 4.5. Note where the value of the measured inventory needs to be entered

4.5 Hillslope Sediment Budget Examples and Inclusion of Uncertainty

The data presented in Table 4.2 can be used to check model coding by comparing model outputs derived with this dataset to the results presented in Table 4.6. It is important to note that Table 4.6 is provided for checking of the central result only and does not include an uncertainty. It is critical that users report their soil erosion estimates with an appropriate uncertainty and that the means by which the uncertainty has been determined is transparent to the reader of reports and subsequent publications.

Uncertainty can be derived from a range of different factors within the sampling and analysis process (Fig. 4.5). The most obvious source of uncertainty is the analytical error originating from the gamma spectrometry measurement (see Chap. 3) which can easily be propagated through to the resulting soil erosion estimate. This is, however, arguably less environmentally relevant than other factors. An alternative

Table 4.5 Microsoft Excel coding for user interface cells (linked to cells described in Table 4.4). Note that until the Microsoft Excel Solver is run using the coding provided below, the numbers in the results cell are arbitrary. Using the 'Solver' command the modelled inventory can be fitted to the measured inventory to allow cell O5 to equal the sample point inventory by changing the constant. Values of h_0 and particle size correction can be changed manually in cells O3-4 for scenario testing

Cell number	Code	Column title and explanation
O2	Enter any start value e.g. 1	Cell where value of C will be calculated by the Solver routine
O3	n/a	Enter particle size correction factor P (enter 1 if no correction required)
O4	n/a	Enter value of h_0
O5	Enter the cell code for column J at the last date in the time series e.g. = J128 would be appropriate for a record with a 127 day extended time series period	This cell value will change depending on the value of the constant C. Initially the number showing will be an artefact of the entry in cell O2 until the solver routine has been run
O7	=SUM(H2:H##) where ## is the row number corresponding to the last time step in the time series	Total erosion based on sum of all values of R (kg m^{-2}). This is the final result but should only be recorded after the solver routing has been run for the worksheet for the specific sample point

Table 4.6 Model output data using the demonstration dataset in Table 4.2 (where $P = 1.61$ and $P' = 0.82$)

Sample ID	PDM soil redistribution amount (no particle size correction) (kg m^{-2})	PDM soil redistribution amount (particle size corrected Taylor et al. 2014) (kg m^{-2})	PDM soil redistribution amount (particle size corrected Yang et al. 2013) (kg m^{-2})
1	−7.9	−3.9	−4.9
2	−2.1	−1.3	−1.3
3	−4.2	−2.3	−2.6
4	−6.7	−3.4	−4.1
5	−4.1	−2.3	−2.5
6	−4.7	−2.6	−2.9
7	−2.8	−1.6	−1.7
8	−2.1	−1.2	−1.3
9	−3.5	−2.0	−2.2
10	−1.4	−0.8	−0.9
11	5.7	4.5	3.2
12	10.5	8.4	6.0

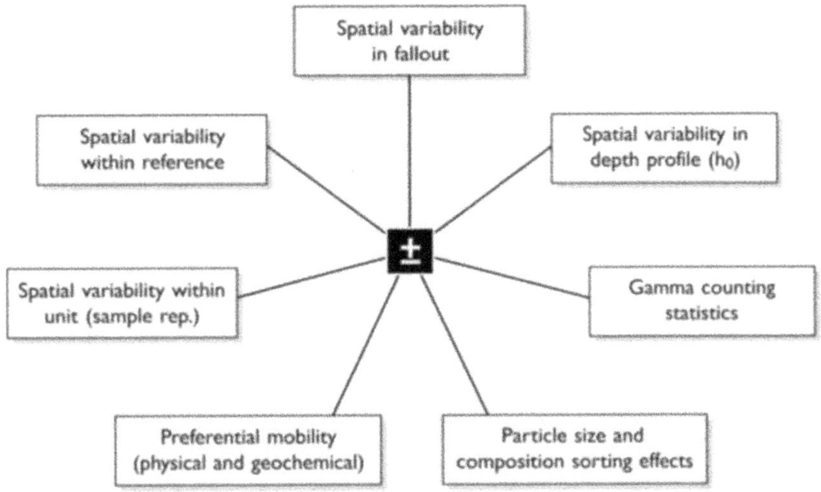

Fig. 4.5 Potential sources of uncertainty in ^{7}Be-based estimates of soil erosion (Taylor et al. 2013)

and more relevant approach is to work with the uncertainty derived from repeated reference inventory measurements. Herein, the coefficient of variation between cores represents a real-world measure of local variability which should be accounted for in erosion estimates. In practice, the model can be run with the upper and lower limits of the reference inventory and the range of outputs used accordingly for each sample point. It should be noted that results from points that differ marginally from the inventory might switch between eroding and depositing depending in reference uncertainty. Similarly, where a spatial integration approach (see Chap. 2) has been adopted in plot sampling, variability in inventory within zones can be considered. Whichever approach is selected, the user must report what was performed with full transparency.

Any approach taken to determine and report uncertainty will have inevitable site-specific and study-specific caveats. Users are encouraged to explore potential sources of uncertainty and undertake suitable sensitivity analysis when running conversion models to convince the scientific and stakeholder communities of data quality and the environmental relevance of results.

Glossary

Profile Distribution Model (PDM) — Simple model to convert ^{7}Be inventory loss into soil erosion amounts by linking inventory change to the depth profile to derive the mass depth of soil loss. This model is applied to single events of short period of time.

Extended time series model — Improved conversion model that permits the effect of radioactive decay on soil erosion rate

	estimates using ^{7}Be over a longer period (e.g. wet season) to be accounted for.
Sensitivity analysis	Process to link uncertainty in the output of a model to different sources of uncertainty in its inputs with the aim of an increased understanding of the relationships between input and output variables.

References

Blake, W. H., Wallbrink, P. J., Wilkinson, S. N., Humphreys, G. S., Doerr, S. H., Shakesby, R. A., et al. (2009). Deriving hillslope sediment budgets in wildfire-affected forests using fallout radionuclide tracers. *Geomorphology, 104,* 105–116.

Blake, W. H., Walling, D. E., & He, Q. (1999). Fallout beryllium-7 as a tracer in soil erosion investigations. *Applied Radiation and Isotopes, 51*(5), 599–605.

Day, P. R. (1965). Particle fractionation and particle-size analysis. In C. A. Black (Ed.), *Methods of soil analysis, Part 1* (pp. 545–567). Madison, Wisconsin: American Society of Agronomy, Inc.

He, Q., & Walling, D. E. (1996). Interpreting particle size effects in the adsorption of Cs-137 and unsupported Pb-210 by mineral soils and sediments. *Journal of Environmental Radioactivity, 30*(2), 117–137.

He, Q., Walling, D. E., & Wallbrink, P. J. (2002). Alternative methods and radionuclides for use in soil-erosion and sedimentation investigations. In F. Zapata (Ed.), *Handbook for the assessment of soil erosion andsedimentation using environmental radionuclides* (pp. 185–215). Dordrecht, Germany: KluwerAcademic Publishers.

Morgan, R. P. C. (2001). A simple approach to soil loss prediction: a revised Morgan-Morgan-Finney model. *CATENA, 44*(4), 305–322.

Porto, P., & Walling, D. E. (2014). Use of ^{7}Be measurements to estimate rates of soil loss from cultivated land: Testing a new approach applicable to individual storm events occurring during an extended period. *Water Resources Research, 50*(10), 8300–8313.

Schuller, P., Walling, D. E., Iroume, A., & Castillo, A. (2010). Use of beryllium-7 to study the effectiveness of woody trash barriers in reducing sediment delivery to streams after forest clearcutting. *Soil & Tillage Research, 110*(1), 143–153.

Taylor A. (2012) The environmental behaviour of beryllium-7 and implications for its use as a sediment tracer. Ph.D. thesis, University of Plymouth, UK.

Taylor, A., Blake, W. H., Smith, H. G., Mabit, L., & Keith-Roach, M. J. (2013). Assumptions and challenges in the use of fallout beryllium-7 as a soil and sediment tracer in river basins. *Earth-Science Reviews, 126,* 85–95.

Taylor, A., Blake, W. H., & Keith-Roach, M. J. (2014). Estimating Be-7 association with soil particle size fractions for erosion and deposition modelling. *Journal of Soils and Sediments, 14*(11), 1886–1893.

Walling, D. E., & He, Q. (1999). Improved models for estimating soil erosion rates from ^{137}Cs measurements. *Journal of Environmental Quality, 28,* 611–622.

Walling, D. E., He, Q., & Appleby, P. G. (2002). Conversion Models for Use in Soil-Erosion, Soil-Redistribution and Sedimentation Investigations. In F. Zapata (Ed.), *Handbook for the assessment of soil erosion and sedimentation using environmental radionuclides* (pp. 111–164). Dordrecht, The Netherlands: Kluwer.

Walling, D. E., He, Q., & Blake, W. (1999). Use of Be-7 and Cs-137 measurements to document short- and medium-term rates of water-induced soil erosion on agricultural land. *Water Resources Research, 35*(12), 3865–3874.

Walling, D. E., Schuller, P., Zhang, Y., & Iroume, A. (2009). Extending the timescale for using beryllium 7 measurements to document soil redistribution by erosion. *Water Resources Research, 45*.

Yang, M.-Y., Walling, D. E., Sun, X.-J., Zhang, F.-B., & Zhang, B. (2013). A wind tunnel experiment to explore the feasibility of using beryllium-7 measurements to estimate soil loss by wind erosion. *Geochimica et Cosmochimica Acta, 114,* 81–93.

Yin, S., Xie, Y., Nearing, M. A., & Wang, C. (2007). Estimation of rainfall erosivity using 5-to 60-minute fixed-interval rainfall data from China. *CATENA, 70*(3), 306–312.

The opinions expressed in this chapter are those of the author(s) and do not necessarily reflect the views of the International Atomic Energy Agency, its Board of Directors, or the countries they represent.

Open Access This chapter is licensed under the terms of the Creative Commons Attribution 3.0 IGO License (https://creativecommons.org/licenses/by/3.0/igo/), which permits use, sharing, adaptation, distribution and reproduction in any medium or format, as long as you give appropriate credit to the International Atomic Energy Agency, provide a link to the Creative Commons licence and indicate if changes were made.

The use of the International Atomic Energy Agency's name, and the use of the International Atomic Energy Agency's logo, shall be subject to a separate written licence agreement between the International Atomic Energy Agency and the user and is not authorized as part of this CC-IGO licence. Note that the link provided above includes additional terms and conditions of the licence.

The images or other third party material in this chapter are included in the chapter's Creative Commons licence, unless indicated otherwise in a credit line to the material. If material is not included in the chapter's Creative Commons licence and your intended use is not permitted by statutory regulation or exceeds the permitted use, you will need to obtain permission directly from the copyright holder.

Chapter 5
Research into Practice—Linking Be-7 Evidence to Land Management Policy Change for Improved Food Security

M. Benmansour, W. H. Blake and L. Mabit

5.1 The Importance of the Short-Term Perspective for Land Management Policy Makers

The short-term perspective on soil redistribution in the landscape offered by ^{7}Be is essential to policy makers with responsibility for developing efficient land management strategies to support food and water security. Soil conservation is vital for enhancement of food production on hillslopes of agroecosystems. Mobilisation of eroded soil and transfer downstream leads to siltation of river channels, lakes and reservoirs, which presents a credible threat to river basin ecosystem service provision and water security. In addition, energy security is threatened by siltation behind hydropower dams. The short-term perspective of this cosmogenic radioisotopic tool means it can provide relatively rapid assessment of very recent changes in practice. Because of its agro-environmental behaviour, this short-lived radioisotope a reliable natural isotopic tracer to assess the effectiveness of recent soil conservation strategies (Mabit et al. 2008; Taylor et al. 2013).

The key challenge that many policy organisations face is obtaining convincing and scientifically sound evidence upon which realistic and wise informed decisions can be based. While many conventional methods are capable of determining on site

M. Benmansour
Centre National de l'Energie, des Sciences et des Techniques Nucléaires (CNESTEN), Rabat, Morocco
e-mail: benmansour64@yahoo.fr

W. H. Blake
School of Geography, Earth and Environmental Sciences, University of Plymouth, Plymouth, UK
e-mail: william.blake@plymouth.ac.uk

L. Mabit (✉)
Soil and Water Management and Crop Nutrition Subprogramme, Joint FAO/IAEA Division of Nuclear Techniques in Food and Agriculture, International Atomic Energy Agency, Vienna, Austria
e-mail: L.Mabit@iaea.org

© International Atomic Energy Agency (IAEA) 2019

L. Mabit and W. Blake (eds.), *Assessing Recent Soil Erosion Rates through the Use of Beryllium-7 (Be-7)*, https://doi.org/10.1007/978-3-030-10982-0_5

erosion rates at a point (e.g. rainfall simulation, erosion pin measurements, survey techniques), the scaling of these measurements to the wider hillslope or landscape unit is challenging. This is especially the case given that equipment associated with many conventional approaches interferes with either farming activities themselves or natural soil erosion processes.

An advantage of ^{7}Be is that this isotopic tracer is integrated into the natural hydrological cycle. Delivered naturally via rainfall to the soil surface and redistributed via erosion processes, it operates without any interference by deployment of measurement equipment. In addition, the method can be applied retrospectively and sampling strategies driven by field observations after an erosion event. The method is not without complications and challenges but if undertaken with care and due attention to assumptions and limitations (as described in Chaps. 1–4), the above advantages offer opportunity to gain critical information on soil and sediment redistribution to support policy decisions on soil conservation and environmental management.

5.2 Linking Nuclear Techniques in Soil Erosion and Conservation to Policy Change

Early studies conducted in the 1990s outlined the principles and opportunity for use of ^{7}Be as a soil erosion and soil redistribution tracer (Wallbrink and Murray 1996; Blake et al. 1999; Walling et al. 1999; Blake et al. 2002). This foundation-laying research has since had a wider impact on soil conservation and environmental management policy through applied studies of the method around the world. This has been facilitated by research development and refinement performed within IAEA's Coordinated Research Projects (CRPs) such as the CRP D1.50.08 "*Assess the effectiveness of soil conservation techniques for sustainable watershed management using fallout radionuclides*" (2002–2008), the CRP D1.20.11 "*Integrated Isotopic Approaches for an Area-wide Precision Conservation to Control the Impacts of Agricultural Practices on Land Degradation and Soil Erosion*" (2009–2013) and more recently through the on-going CRP D1.50.17 "*Nuclear Techniques for a Better Understanding of the Impact of Climate Change on Soil Erosion in Upland Agro-ecosystems*" (2016–2021). Following the CRPs activities, several guidelines were produced by the SWMCN Subprogramme of the Joint FAO/IAEA Division of Nuclear Techniques in Food and Agriculture (e.g. IAEA 2014; Mabit et al. 2018).

Impact on policy has been further extended worldwide to several IAEA Member States and FAO Member Countries by methodology transfer and capacity-building activities within IAEA national and/or regional technical cooperation projects (e.g. RAF5075 "*Enhancing Regional Capacities for Assessing Soil Erosion and the Efficiency of Agricultural Soil Conservation Strategies through Fallout Radionuclides in Africa*") which includes targeted training courses on the use of ^{7}Be as a tracer to support soil conservation policy by leading researchers. As a result of these outreach and science-policy impact activities, application of this specific methodology has been extended to a range of soil conservation challenges, in collaboration with national and international organisations responsible for environmental management policy to support food and water security or management of forest resources.

For example, Blake et al. (2009) undertook a landscape sediment budget approach with ^{7}Be alongside ^{137}Cs and ^{210}Pb to evaluate the post-fire loss of topsoil and nutrients from forest affected by wildfire to a major water supply reservoir in Australia, in collaboration with the government agency responsible for developing policy on fire management for water quality and quantity. Under future climates, wildfire will exacerbate threats to water security. The applied fallout radionuclide (FRN) research demonstrated that burning of surface vegetation can invigorate hillslope hydrological response with marked increases in sediment and nutrient delivery to river networks and reservoirs. Negative water quality effects include for example high turbidity, toxic algal blooms and fish kills with implications for water supply at critical times in the water year. Through quantifying post-fire runoff and nutrient yield processes, the research informed catchment management decisions, policies and water resource risk assessment in Australia. Similarly, researchers in Chile, applied ^{7}Be using a transect approach (Schuller et al. 2006) for evaluation of the impact of forest harvesting on soil erosion fine sediment yields and for contribution to improve the effectiveness of mitigation measures before, and after clear-cut operations and during the establishment of the new plantations. This work demonstrated the benefit of using ^{7}Be to document the effects of recent forest clearance upon slope soils in a timber harvesting operation. The short-term perspective of ^{7}Be meant the use of mitigation strategies could be assessed and, in this case, a high sediment delivery ratio indicated to policymakers the need for further soil conservation strategies.

5.3 Example Impact Case Study: Support Provided by The ^{7}Be Technique to Shape Soil Conservation Policy in Morocco

5.3.1 Soil Erosion and Conservation Policy Challenges in Morocco

Soil erosion is the main land degradation process in Morocco, which is affecting up to 40% of its land area. About two million hectares of Moroccan agricultural lands are affected by soil erosion by water (Dahan et al. 2012). On average, soil erosion ranges from 5 to 20 t ha^{-1} yr^{-1}, but exceeds these magnitudes in northern and north-western basins. For example, in the pre-Rif hills region, soil erosion exceeds 50 t ha^{-1} yr^{-1}(HCEFLCD 2017). Out of a total area of 20 million ha of watershed in Morocco, 50% is estimated to be subjected to very high erosion risks with a yearly soil loss of around 100 million tons. This annual soil loss leads to a reduction of 75 million m^{3} of downstream dam water storage capacity. Each year about 0.5% of the country's reservoir storage capacity is lost. These soil erosion and conservation challenges require both spatially representative data on soil erosion amounts and sediment delivery ratios. The highly dynamic nature of land use change and implementation of new conservation policies, however, means that long- or medium-term FRN approaches are not recommended to deliver the data required by policymakers on the extent of the problem and the level of success in conservation approaches applied (Mabit et al. 2008). Consequently, researchers in Morocco have been using the ^{7}Be methodologies outlined in this book to tackle these challenges and the obtained data have been used to assist development of new soil conservation policy in a wide range of settings.

5.3.2 From the Laboratory to the Field: The Approach

FRN methodologies have been successfully used by the "Centre National de l'Energie, des Sciences et des Techniques Nucléaires" (CNESTEN) in many cultivated areas of Morocco for supporting soil conservation strategies tested or implemented by the government. The research has shaped soil conservation policy with the collaboration of "Institut National de Recherche Agronomique (INRA) from the Ministry of Agriculture and Maritime Fisheries and "Centre de Recherche Forestière" (CRF) from the "Haut Commissariat des Eaux et la Lutte contre la Désertification" (HCEFLCD). Fallout ^{7}Be was applied in combination with ^{137}Cs and ^{210}Pb to estimate soil erosion in agricultural or forest fields under different land uses in the regions of Rabat, Tétouan, Casablanca and Fes (Fig. 5.1) both before and after implementation of recommended soil conservation approaches. The transect sampling strategy was generally adopted (see Chap. 2) and FRNs were determined at CNESTEN Laboratory using gamma detectors (see Chap. 3). Key studies where the ^{7}Be approach has been integrated into policy-development research:

(1) In semi-arid and Mediterranean climatic areas (Rabat and Tétouan regions), a no-till technique was tested as a soil conservation practice and compared to conventional tillage in the experimental sites "Marchouch" and " Herchane" of the region of Rabat and Tétouan respectively;
(2) In semi-arid climatic areas (Casablanca-Settat region), ^{7}Be was used to evaluate the efficiency of conservation practices involving Atriplex plantations (implemented based on results from other FRNs) and, cereal and fruit plantations effectiveness were assessed within the "Oued Mellah" watershed management programme;
(3) In variable climatic areas from cold-humid to semi-arid climate (Fes region), the efficiency of soil conservation practices based on the use of Allepo pine trees combined with dry stones and gabion and also fruit plantations was assessed within the framework of the "Allal Fassi" watershed management programme.

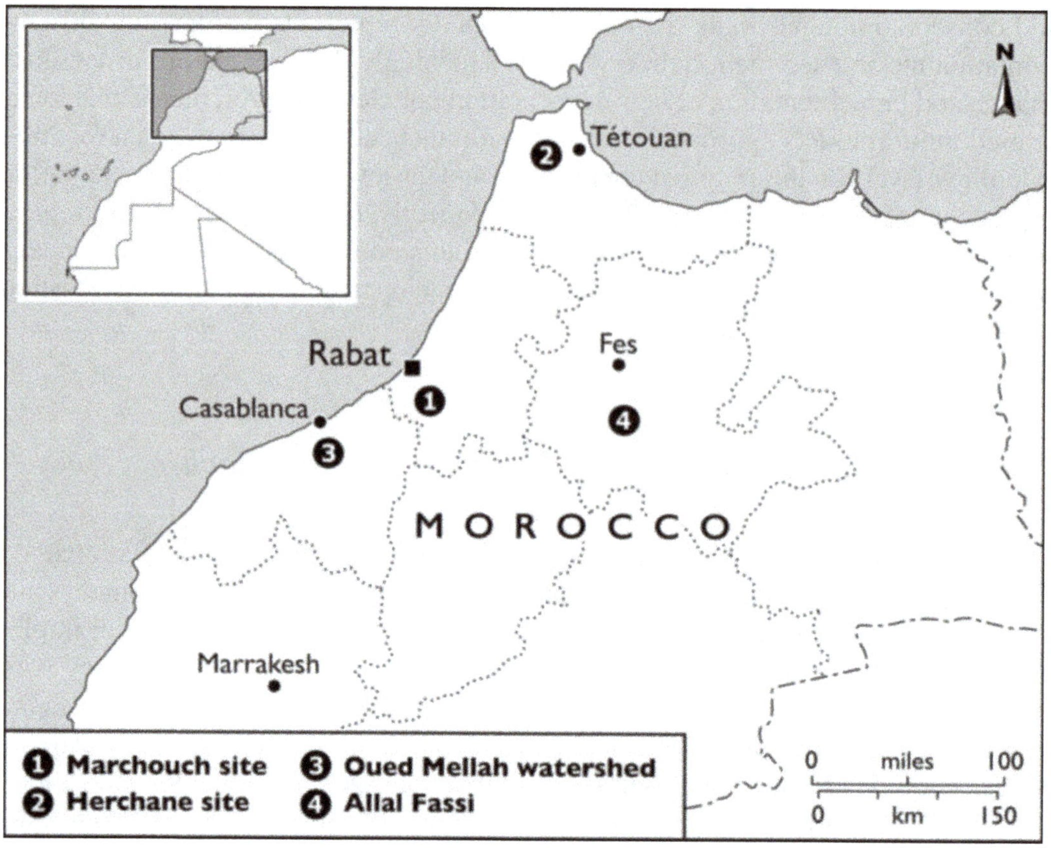

Fig. 5.1 Locations of the study areas = (1) Marchouch site (region of Rabat), (2) Herchane site (region of Tétouan), (3) Oued Mellah watershed (region of Casablanca) and (4) Allal Fassi (region of Fes)

5.3.3 Key Findings and Policy Impacts from ^{7}Be Application in Morocco

Previous investigations using other FRNs (^{137}Cs or ^{210}Pb$_{ex}$) allowed establishment of the long-term soil erosion rates of the three investigated regions, which ranged from 8 to 58 t ha^{-1} yr^{-1} (Benmansour et al. 2013; Benmansour et al. 2016; Yassin et al. 2017). The net soil erosion rates appeared to be closely related to the rainfall, slope and the past land use. The sediment delivery ratio in all areas is generally high reaching 80 to 100% which led to implementation of soil conservation strategies to keep fertile soil material on the hillslopes. The short-term perspective of ^{7}Be enabled the effectiveness of these soil conservation strategies to be assessed and underpinned their wider adoption in the landscape. Figure 5.2 reports the mean values of short-term erosion amounts estimated at different study sites and corresponding to fields under conventional tillage or without management and those corresponding to fields under soil conservation practices.

Soil redistribution data from ^{7}Be measurements indicated that soil loss had been reduced significantly under no-till as compared to conventional tillage in Rabat and Tétouan regions. Indeed, soil erosion rates were lowered by 50% for the Marchouch site (Fig. 5.2a) and by 40% for the Harchane site (Fig. 5.2b). At the Oued Mellah watershed, the results highlighted that high density Atriplex plantations reduced soil loss by approximately 57 to 80% compared to Atriplex plantations with low density while for the site under fruit plantations and cereals, soil erosion has been decreased by 58% compared to bare soils (Fig. 5.2c). For Allal Fassi Watershed, erosion was lowered by 54% for soil under young Allepo pine, dry stones and gabions compared to soil under only Allepo pine and dry stones and by 51% for agricultural fields under cereals and fruit plantation as compared to bare soils (Fig. 5.2d). These findings have emphasised the need for wider adoption of these proven soil conservation approaches in the Moroccan landscape. Indeed, the no-till practice tested by Moroccan INRA researchers in their experimental sites is being applied and transferred by the Ministry of Agriculture to agricultural farms that have similar agro-environmental conditions as the experimental stations of INRA. The demonstrated high performance of soil conservation practices tested by HCEFLCD in the investigated watersheds means they will be adopted at a wide scale.

The relevant results derived from the application of these FRNs for short and longer-term erosion rates in Morocco were presented jointly by CNESTEN and HCEFLCD at the 12th Session of *United Nation Convention for Combating Desertification, COP12*, Turkey, October 2015. The impact of this research on policy was emphasised by a lead policy maker, the Head of the Forestry Research Centre of Morocco's High Commission of Forest and Water and Combating Desertification (HCEFLCD). He spoke about the role of isotopic techniques in tackling the effects of climate change, particularly drought leading to a reduction of up to 75% in grain yields: "*Using isotopic techniques, we were able to accurately assess soil erosion and the effectiveness of soil conservation practices and make concrete recommendations to policy makers… These all delivered real change for people who rely on the land for their livelihoods.*" (IAEA 2016).

5.4 Future Trends and Opportunities in Using ^{7}Be

The above science-to-policy examples in Morocco illustrate the fundamental role that the short-term perspective on soil redistribution rates provided by ^{7}Be can play in the development and implementation of soil erosion policy. The need for such information has never been so crucial. Against threats from soil erosion, it is predicted that global food production must increase by 70% to feed the projected growth of the world's population (from 7 to 9 billion by 2050) (FAO 2017). Extreme weather (wet and/or dry) events may become more frequent with changing global climate with impacts on soil quality and erodibility. Developing community resilience to such recurrent climatic events demands 'win–win' systems of soil conservation that permit enhancement of land productivity while protecting ecosystem services in the

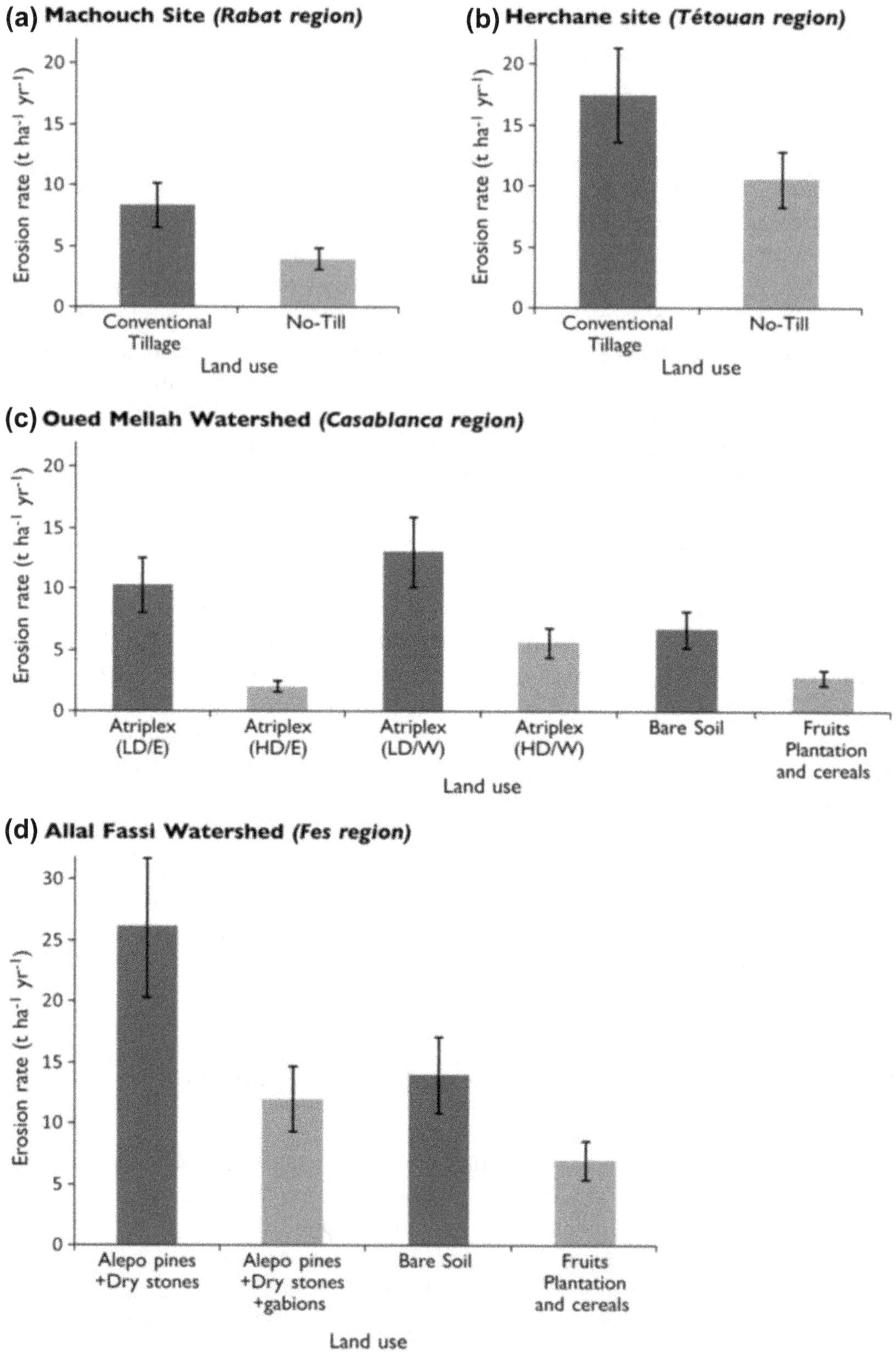

Fig. 5.2 Short soil erosion rates under different land uses associated with the study sites: Marchouch (**a**), Herchane (**b**), OuedMellah (**c**) and AllalFassi (**d**). *LD* Low density; *HD* High density; *E* East exposure; *W* West exposure

wider river basin. Key information and guidance provided by nuclear and isotopic tools such as the ^{7}Be technique provides valuable support to convince policy makers and farmers to adopt and promote widely existing effective climate smart soil conservation strategies.

References

Benmansour, M., Mabit, L., Nouira, A., Moussadek, R., Bouksirate, H., Duchemin, M., et al. (2013). Assessment of soil erosion and deposition rates in a Moroccan agricultural field using fallout ^{137}Cs and ^{210}Pb$_{ex}$. *Journal of Environmental Radioactivity, 115*, 97–106.

Benmansour, M., Mabit, L., Moussadek, R., Yassin, M., Nouira, A., Zouagui, A., et al. (2016). Effectiveness of soil conservation strategies on erosion in Morocco. In *Geophysical Research Abstracts* (Vol. 18), European Geosciences Union—General Assembly 2016. Abstract EGU2016-2174.

Blake, W. H., Walling, D. E., & He, Q. (1999). Fallout beryllium-7 as a tracer in soil erosion investigations. *Applied Radiation and Isotopes, 51*(5), 599–605.

Blake, W. H., Walling, D. E., & He, Q. (2002). Using cosmogenic beryllium-7 as a tracer in sediment budget investigations. *GeografiskaAnnaler Series a-Physical Geography, 84A*(2), 89–102.

Blake, W. H., Wallbrink, P. J., Wilkinson, S. N., Humphreys, G. S., Doerr, S. H., Shakesby, R. A., et al. (2009). Deriving hillslope sediment budgets in wildfire-affected forests using fallout radionuclide tracers. *Geomorphology, 104*(3–4), 105–116.

Dahan, R., Boughala, M., Mrabet, R., Laamari, A., Balaghi, R., & Lajouad, L. (2012). *A review of available knowledge on land degradation in Morocco* (p. 48). Oasis Country Report 2, International Center for Agricultural Research in the Dry Area(ICARDA).

FAO. (2017). The future of food and agriculture—Trends and challenges (163 p). FAO publication.

Haut Commissariat des Eaux et Forêts et à la Lutte contre la Désertification (HCEFLCD). (2017): http://www.eauxetforets.gov.ma.

IAEA. (2014). *Guidelines for using Fallout radionuclides to assess erosion and effectiveness of soil conservation strategies*. IAEA-TECDOC-1741. Vienna, Austria: International Atomic Energy Agency Publication.

IAEA. (2016). *17 Ways to change the world: IAEA promotes the role of nuclear technologies in sustainable development at European development days*. Available online at https://www.iaea.org/newscenter/news/17-ways-to-change-the-world-iaea-promotes-the-role-of-nuclear-technologies-in-sustainable-development-at-european-development-days. Accessed November 2017.

Mabit, L., Benmansour, M., & Walling, D. E. (2008). Comparative advantages and limitations of the fallout radionuclides ^{137}Cs, ^{210}Pb$_{ex}$ and ^{7}Be for assessing soil erosion and sedimentation. *Journal of Environmental Radioactivity, 99*(12), 1799–1807.

Mabit, L., Bernard, C., Lee Zhi Yi, A., Fulajtar, L., Dercon, G., Zaman, M., et al. (2018). Promoting the use of isotopic techniques to combat soil erosion: An overview of the key role played by the SWMCN Subprogramme of the Joint FAO/IAEA Division over the last 20 years. *Land Degradation & Development, 29*, 3077–3091.

Schuller, P., Iroume, A., Walling, D. E., Mancilla, H. B., Castillo, A., & Trumper, R. E. (2006). Use of beryllium-7 to document soil redistribution following forest harvest operations. *Journal of Environmental Quality, 35*(5), 1756–1763.

Taylor, A., Blake, W. H., Smith, H. G., Mabit, L., & Keith-Roach, M. J. (2013). Assumptions and challenges in the use of fallout beryllium-7 as a soil and sediment tracer in river basins. *Earth-Science Reviews, 126*, 85–95.

Wallbrink, P. J., & Murray, A. S. (1996). Distribution and variability of Be-7 in soils under different surface cover conditions and its potential for describing soil redistribution processes. *Water Resources Research, 32*(2), 467–476.

Walling, D. E., He, Q., & Blake, W. (1999). Use of Be-7 and Cs-137 measurements to document short- and medium-term rates of water-induced soil erosion on agricultural land. *Water Resources Research, 35*(12), 3865–3874.

Yassin, M., Benmansour, M., Chikhaoui, M., Ismaili Alaoui, F. Z., El Bahi, S., Babaou,Y., et al. (2017). Contribution à l'évaluation de l'impact de l'aménagement des bassins versants de l'Oued Mellah et Allal El Fassi au Maroc. *Annales de la Recherche Forestière au Maroc, 44*, 79–96.

The opinions expressed in this chapter are those of the author(s) and do not necessarily reflect the views of the International Atomic Energy Agency, its Board of Directors, or the countries they represent.

Open Access This chapter is licensed under the terms of the Creative Commons Attribution 3.0 IGO License (https://creativecommons.org/licenses/by/3.0/igo/), which permits use, sharing, adaptation, distribution and reproduction in any medium or format, as long as you give appropriate credit to the International Atomic Energy Agency, provide a link to the Creative Commons licence and indicate if changes were made.

The use of the International Atomic Energy Agency's name, and the use of the International Atomic Energy Agency's logo, shall be subject to a separate written licence agreement between the International Atomic Energy Agency and the user and is not authorized as part of this CC-IGO licence. Note that the link provided above includes additional terms and conditions of the licence.

The images or other third party material in this chapter are included in the chapter's Creative Commons licence, unless indicated otherwise in a credit line to the material. If material is not included in the chapter's Creative Commons licence and your intended use is not permitted by statutory regulation or exceeds the permitted use, you will need to obtain permission directly from the copyright holder.